Earth and Environmental Sciences Library

Earth and Environmental Sciences Library (EESL) is a multidisciplinary book series focusing on innovative approaches and solid reviews to strengthen the role of the Earth and Environmental Sciences communities, while also providing sound guidance for stakeholders, decision-makers, policymakers, international organizations, and NGOs.

Topics of interest include oceanography, the marine environment, atmospheric sciences, hydrology and soil sciences, geophysics and geology, agriculture, environmental pollution, remote sensing, climate change, water resources, and natural resources management. In pursuit of these topics, the Earth Sciences and Environmental Sciences communities are invited to share their knowledge and expertise in the form of edited books, monographs, and conference proceedings.

More information about this series at http://www.springer.com/series/16637

Aleksander V. Semenov · Igor A. Sokolov
Editors

Sustainable Development: Society, Ecology, Economy

Proceedings of the XVth International Scientific Conference 2019, 28 March 2019, Moscow Witte University

Springer

Editors
Aleksander V. Semenov
Moscow Witte University
Moscow, Russia

Igor A. Sokolov
Federal Research Center Informatics and Management of the Russian Academy of Sciences
Moscow Witte University
Moscow, Russia

ISSN 2730-6674 ISSN 2730-6682 (electronic)
Earth and Environmental Sciences Library
ISBN 978-3-030-73109-0 ISBN 978-3-030-73110-6 (eBook)
https://doi.org/10.1007/978-3-030-73110-6

This Springer imprint is published by the registered company Springer Nature Switzerland AG
The registered company address is: Gewerbestrasse 11, 6330 Cham, Switzerland

Preface

At the present, the community of scientists and practitioners has a significant and continuously increasing knowledge potential in the field of sustainable development of society. From the point of view of the world community, the unconditional flagship and leader of research and implementation of the requirements of sustainable development is the UN, which has been first to set this task in the last millennium and still correcting the well-known aims and objectives of sustainable development of human civilization, however already in modern conditions.

Any system (household, society, economy, state, nature, etc.) may develop sustainably both along an upward trend and along a downward trajectory. In the first case, there is a steady growth, or steadily progressive dynamics and direction of this development. In the second case (downward trajectory), there is a steady recession, or steadily regressive dynamics and direction of a certain system. At the same time, the steady growth and steady recession of the system does not mean the strictly linear nature of the corresponding dynamics at all; there may also be non-linear, precisely trend, or "wave-like" characteristics of the named forms of sustainability. In other words, there may be some deceleration and acceleration of the rates of sustainable growth or a sustainable recession in the parameters and indicators of the system. However, at the same time, not a single "point" of the general trend is characterized by reciprocating or turbulent dynamics.

Thus, as a result of a qualitative analysis of the phenomenon of "sustainable development", the essence of sustainable development can be defined as the development of a system based on the dynamic balance of the relevant conditions and factors.

Obviously, sustainable development can be both progressive and regressive. Sustainable progressive development is always advancing, namely the creative development of the system, carried out on the basis of a dynamic balance of conditions and factors that are simultaneously improving. Such an interpretation aims the subjects of management of various systems at discovery, study of the ontology and skillfull combination of external and internal conditions and factors in order to ensure the actual progress of a particular system. In a similar way, sustainable regressive development is always anti-progressive and anti-creative development of the system, carried out on the basis of dynamic equilibrium, but synchronously collapsing, dysfunctionally realized conditions and factors. Sustainable neutral development may also be defined as a development that is the result of a synchronous interaction of conditions and factors that retain their qualitative properties and quantitative characteristics unchanged.

In accordance with the cyclical development of any system, trends for sustainable progressive and sustainable regressive development periodically arise.

As evidenced by many facts of reality, in the modern development of man, society (civilization), economy and nature, a trend toward systemic destruction of seemingly "equilibrium" relations and interactions begins to prevail. In this regard, a number of systemic questions arise that also require systemic answers and which are absent yet in an integral form in the current program documents, as well as in the scientific and special literature.

The triad of the international concept of sustainable development is made up of economic, social, and environmental processes.

The modern system for measuring socio-economic processes is imperfect. There is a growing understanding and recognition of the fact that GDP is not an ideal indicator for measuring human well-being, since it does not cover various social processes, changes in the environment, some phenomena that are commonly called "sustainability" of development. GDP growth can occur against the background of degradation of human, physical, and natural capital. Taking into consideration of this circumstance, quantitative indicators, either cost indicators (GDP, etc.) or physical volumes (oil, gas, metals, etc.), are not adequate for a sustainable development system. One of the directions for improving the system of measuring socio-economic processes is the need to shift the emphasis in the system of indicators from measuring production to measuring human well-being. At the same time, the measurement of well-being should be considered in the context of ensuring sustainable development. The fundamental point is that costs for human development are added and damage from environmental pollution and depletion of natural resources are deducted from the organization's profit. Thus, a new economic reality is being created, in which the emphasis is on the continuous growth of the well-being of each human and the improvement of the state of his natural environment.

In recent years, one of the most important strategic factors in ensuring national and global security has become the quality of life of the population with a correlation to human well-being. Recent studies show that this relatively new and very broad concept characterizes not only the basic living conditions of people, but also the efficiency of their life, as well as people's own attitude to these factors. The latter circumstance is especially important, since it is precisely this that is the main reason for the increase in the level of social tension, which is observed today in many countries and regions of the world.

Currently, the problem of improving the population's quality of life should not be regarded only as one of the issues of the socio-economic development of a particular country. This is a global problem of a higher level—ensuring national and global security. It should constantly be in the spotlight of international organizations, political and public leaders of states, as well as the world scientific and educational community.

At the present, the world community has not yet developed a sufficiently general approach to defining the concept of quality of life. Therefore, the generally accepted international methodology for measuring and quantifying the quality of life is practically absent today. For example, Doctor of Medical Sciences, Prof. I. A. Gundarov believes that the quality of life is a measure of how people's living conditions correspond to their needs and capabilities.

In the international practice of sociological research aimed at a comparative analysis of the life of the population in different countries of the world, instead of the concept of "population's quality of life", the term "population's standard of living" is often used, and less often—the broader concept of "welfare of the population". At the same time, the population's standard of living is considered as a certain integral characteristic of the degree of satisfaction of population's material and spiritual needs with the mass of goods and services that are used by them in a particular period of time, as a rule, during the current year. It is important to note that in order to assess the population's quality of life, a list was approved at the UN level, including the following indicators:

- fertility, mortality, and life expectancy;
- sanitary and hygienic conditions;
- quantity of food consumed;
- housing situation;
- quality of education and culture;
- employment level, working conditions;
- balance of income and expenditure;
- prices for consumers;
- the state of the transport infrastructure;
- recreational system;
- the level of institutional care;
- human rights and freedoms.

Analysis of this list shows that it is basic, however incomplete.

The specificity of the problem of assessing the quality of life is that it is complex and therefore requires taking into account both material and non-material factors, which are rather difficult to formalize and quantify. Indeed, even the term "quality of life" itself is a complex concept with a rather contradictory content, since different cultures have different ideas about what is good. Therefore, when conducting research on the population's quality of life in various countries of the world, currently they use their own methods.

The main difficulties in solving this problem are to develop a constructive approach that would allow structuring this problem in such a way that its main components, factors and indicators could be used in the practice of strategic planning and management, without losing the main humanitarian aim of this management, which is to improve the quality of the population itself, and not just the conditions of its existence. From the standpoint of this approach, a concept is proposed for a comprehensive solution to the problem of assessing the quality of life of the population based on the definition, measurement, and simultaneous consideration of the following three groups of integral indicators:

(1) the quality of the human environment, which is determined by both natural and social factors. It is a combination of mainly material and socio-economic aspects of this environment;
(2) people's attitude to the conditions of their existence and the degree of their satisfaction with these conditions. These are non-material aspects of the quality of life, mainly spiritual and psychological ones;
(3) the quality of the population itself, which is characterized by its viability and capacity for work in the current living conditions.

The quality of life is the most important indicator of sustainable development of society, the value of which goes far beyond the socio-economic sphere. This factor largely determines the aims of socio-economic policy, the psychological atmosphere in society and the level of its social stability, the possibility of social cataclysms.

The concept of sustainable development has determined the ways of developing environmental safety, reaching the integration of the economy and ecology and realizing the need for a different approach in the relationship of man and society with nature. The ecological problem, from the point of view of its genesis, reproduces the stages of development of the contradiction between nature and society, as the development of any contradiction, as a transition from unity, balance to disharmony and confrontation. Human monopoly has become unprecedented, and therefore environmental crises are inevitable.

One of the main issues of modern society, guaranteeing the well-being of future generations, is the rejection of the aggressive, wasteful exploitation of natural resources, without which the existence of mankind is unthinkable. Therefore, special attention is required to the problem of rational use of natural resources for the benefit of society and their restoration within an acceptable period of time. The issue of waste disposal is also no less relevant in modern conditions. Despite the numerous proposed solutions for waste disposal, this problem has not been solved at the system level. One of the directions for solving the problem is seen on the basis of the approach of the waste matter decomposition into nanosized particles and the formation of a commercial product substance from them, presented at the XV International Scientific Conference "Sustainable Development: Society, Ecology, Economics" held at Moscow Witte University in March 2019.

The problems of sustainable development of society mentioned above emerge from a man and, thus, a man must solve them through awareness of his place and role in the ecology of the planet. Therefore, it is necessary to pay the greatest attention to environmental upbringing and education, the formation of environmental ethics, morality, virtue of the whole society. The transformation of environmental education and enlightenment into an integral part of the education program of upbringing and educational institutions is an extremely urgent task for all regions and countries of the world.

In the collection of the conference materials "Sustainable Development: Society, Ecology, Economy" held at the Moscow Witte University, various aspects of social, environmental, and economic processes that make up the triad of the international concept of sustainable development are highlighted. Conceptual priorities and specific aims, quantitative indicators and activities aimed at reaching the goals in the framework of the concept of sustainable development are considered in the materials. The international experience of reaching social and economic effects, ecologization of various projects on the basis of modern achievements of science and technology are presented. A significant attention is paid to the problem of improving human well-being, measured in the context of sustainable development. Within the framework of the conference theme, the participants presented the research results that determine the ways to solve the identified problems and achieve significant results for individuals and society as a whole.

Moscow, Russia

Aleksander V. Semenov
Rector of Moscow Witte University

Contents

Modern Environmental and Climatic Problems and Sustainable Development of Society

The Conception of Power Generation for Sustainable Social Development

Aleksandr B. Shapovalov

Abstract On the basis of distributed autonomous power generation and power generation based on nano-decomposition, the concept of power generation for sustainable social development is presented.

Keywords Society · Distributed power generation · Environmental safety · Carbon-containing fuels · Conflict potential · Nano-decomposition · Nanoparticles

The life of a society as a continuous transformation of matter is determined by an energy, which is [1] a general quantitative measure of the movement and interaction for all types of matter.

Numerous assessments [2] of the power consumption dynamics of a society (Fig. 1a) show a constant increase. The exponential growth of society (Fig. 1b) corresponded to the growth of power consumption per individual. However, power consumption per individual in recent years (Fig. 1b) is approximately constant.

The disappearance of the matter and energy transformation growth per individual is a sign of stagnation. A decrease in the growth rate of society will lead to the cessation of its exponential growth and, therefore, to the decrease in power consumption per individual by the end of this century [2]. A decrease in the matter and energy transformation means a degradation of a society.

Globally, the dynamics of biological evolution [3] doesn't depend on the society. Hence, the causes of society stagnation are anthropogenic limitations of resources, both in the form of energy and its structural form.

The monopolization of power generation within the framework of the dominant market paradigm hinders the sustainable development of society: the development of power generation and fuels versus the deriving of profit from this by monopolies. Precisely this forms the natural conflict potential in modern society [4]: between resources and claims on them. The monopolization of power generation, supporting the "status quo", contributes to the growth of conflict potential in accordance with

A. B. Shapovalov (✉)
Moscow Witte University, 12, 2nd Kozhukhovsky pas., Bldg. 1, Moscow 115432, Russia
e-mail: shapovalov-ab@yandex.ru

A. V. Semenov and I. A. Sokolov (eds.), *Sustainable Development: Society, Ecology, Economy*, Earth and Environmental Sciences Library,
https://doi.org/10.1007/978-3-030-73110-6_1

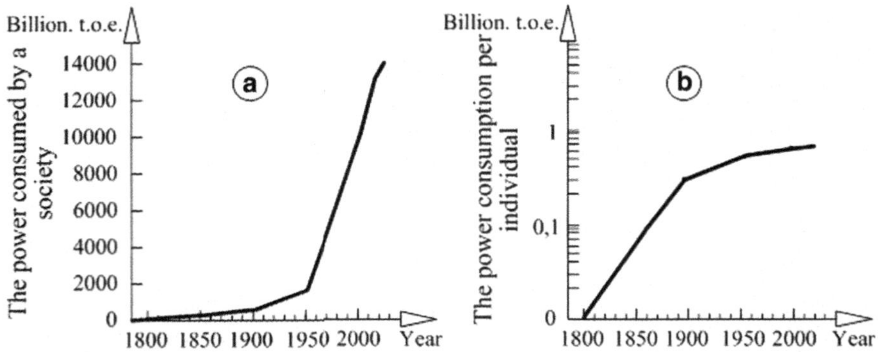

Fig. 1 The power consumption dynamics of a society

the growth of power consumption by the society, leading to its global destabilization [5, 6] setting it on the brink of disaster.

Despite the variety of energy carriers, fossil carbon-containing fuels (CCF) dominate and provide over 86% of modern power generation.

The hopes [7] for alternative energy sources are doubtful. Indeed, any natural factors may be transformed into energy: solar radiation, wind, atmospheric electricity, water movement, geothermal processes, etc.

However, the hydropower resource is, in fact, exhausted. Optimistic estimates of the renewable sources contribution to crude energy production are about 3%. Saturation of the economically justified market for these technologies is noted [7], excluding developing countries (China, Brazil, India, etc.). Thus, [7] the potential of renewable energy sources is almost exhausted.

The contribution of atomic power generation [7.8] to crude energy is about 4.4%. Fossil fuel [7] uranium isotope with a mass of ^{235}U is contained in natural uranium up to 0.7% with a resource for 50-80 years. Even a possible increase in the use of natural uranium [7.8] up to 80 times due to the ^{238}U uranium isotope in expensive and complex breeder reactors (breeders), does not make atomic power generation a source of ecologically safe and cheap energy.

The basic trend [8, 9] is the demonopolization of power generation due to the *rapid growth of distributed autonomous sources of* power *generation.*

There are the following advantages [9, 10] of distributed autonomous power generation: low capital costs, quick commissioning, relatively simple maintenance and environmental safety. There are also no expenses on infrastructure and its operation: there are no power lines, substations, heating networks, etc. The proximity of power generation and consumers eliminates infrastructure losses and increases the reliability of power supply.

The choice of CCF for power generation is determined by numerous factors: the method of exothermic oxidation, physicomechanical properties of CCF, logistic parameters, economic and socio-political indicators. Autonomous power generation is primarily focused on the availability of local CCFs (plant, peat, etc.). The trend

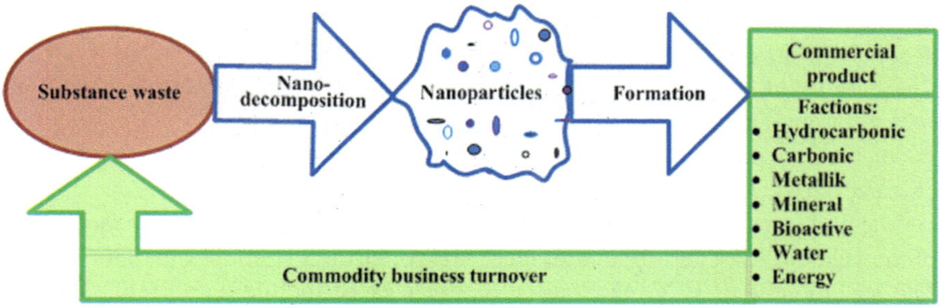

Fig. 2 Recycling by nano-decomposition

[10] of stabilization and sublimation of the local CCFs energy potential significantly increases the efficiency and environmental safety of power generation, leading to the replacement of traditional boiler and gaseous fuels.

Without diminishing the importance of traditional CCFs, local energy-efficient and environmentally friendly CCFs are the basis for the development of autonomous power generation. However, contributing to demonopolization, local CCFs to a greater or lesser extent bear the "tribal seal" of monopolization. Nevertheless, power generation based on local fuels leads to a change in the balance of power consumption, demonopolizing power generation.

However, this is not the only moment: the existence and development of society is due to the continuous transformation of more than 140 million (Chemical Abstracts Service, CAS for 2018) dominant carbon bearing substances.

Carbon bearing substances that lose their expediency (economic, social, etc.) go into the waste category, as a rule, leading to degradation of the biosphere [6]. It is logical [10, 11] *to use the energy potential of carbon bearing substances of waste.*

The nature of solid municipal waste (SMW) defines them as *a natural local fuels*, spared from the "tribal seal" of monopolization. Today, the energy potential of SMW of an individual [11] ensures its household power consumption and, therefore, energy security, ultimately increasing the sustainability of society.

Thus, the complex of local CCFs ensures the *demonopolization of power generation*, at least at the household level (up to 60% of primary energy production).

The heterogeneous nature of CCF (including SMW) during the conversion of their substances causes a variety of secondary synthesis reactions leading to the forming of harmful emissions, including the class of persistent toxic substances (POPs). This is a fundamental obstacle to eco-friendly energy generation.

The only possibility of ecologically safe power generation is realized [6, 10, 11] in the "Shapovalov nano-decomposition model" (Fig. 2.) which was honoured by an international award: *"decomposition of waste material into nanosized particles and the forming of a commercial product substance from them".*

The marketed product in the form of an energy fraction is presented as the activated nanoparticles of the {CnHm} and {C} type [11] with increased reactivity, the energy potential of which can be realized directly or converted into various CCFs.

Nano-decomposition with a rate of temperature rise of $10^3 \div 10^5$ deg/s provides decomposition of practically one compound and eliminates secondary synthesis. Compared with other thermal recycling methods [12], nano-decomposition is the only environmentally friendly.

Such an approach assumes different CCFs variations. In other words, it is characterized by the universality.

The fuel formed by nano-decomposition [11] in the form of activated nanoparticles of the {CnHm} and {C} type is an ideal fuel for electrochemical generators (ECG). The specificities of ECG-generators are the lack of mobile parts, reliability, unit efficiency up to 65% and unique adaptation to outer loads.

Thus, there is a trend in the society for *the rapid growth of distributed autonomous sources of power generation on the basis of stabilization and sublimation of the CCF energy potential. This leads to the demonopolization of power generation.* The culmination of this trend is power generation based on nano-decomposition of waste, *which not only changes the balance of fuels consumed*, but also ***naturally providing an independent household power consumption of an individual in society.*** As a result, at the individual level, a non-volatile self-regulating cell forms, which becomes the basis of the society.

In that case, the contradictions between resources and claims on them are decreased, and, therefore, *the conflict potential in society is also reduced*. Moreover, this approach [11] leads to *corresponding changes in the infrastructure of society. In other words, that leads to its transformation into new forms.*

As a result, the concept of power generation for sustainable social development is:

> distributed autonomous power generation of society on the basis of energy independence and self-regulation of the individual.

The research was conducted on our own initiative and at our own expense.

References

1. Prokhorov AMM (ed) (1983) Physical encyclopedic dictionary. Sovietskaya Encyclopedia, 928 p (In Russian)
2. Petin VG, Belkina SV (2012) Thermodynamic aspects of biosphere sustainability and global environmental problems. Interdiscipl J Basic Appl Sci “Biosfera” 4(1):8–15 (In Russian)
3. Futuyma DJ (2010) Evolutionary theory. In: Muhlenbein MP (ed) Human evolutionary biology. Cambridge University Press, Cambridge, pp 3–16
4. Dahrendorf R (1994) Elements of the social conflict theory. Sociological studies. Russian Academy of Sciences, vol 5, pp 142–147 pp (In Russian)
5. Yushkov IV (2013) On the role of energy resources in international conflicts of the future. Bull Moscow Univ Ser 12: Polit Sci 2:28–37 (In Russian)
6. Lazarev GE (2019) Environmental aspects of sustainable development. Sustainable development: society, ecology, economy. In: in 2 parts/under the ed. of Semenov AV, Malyshev NG, Rudenko Yu S (ed) Materials of the XV international scientific conference; PEIHE “Moscow Witte University”, pp 1–748, pp 97–105 (In Russian)

7. Shapovalov AB (2017) On the issue of alternative power generation. In: 2 parts/under the ed. of Semenov AV, Malyshev NG, Rudenko Yu S (ed) Environmental and nature protection problems of modern society and ways to solve them: materials of the XIII international scientific conference; PEIHE "Moscow Witte University", Part 1 .- 504, pp 127–134 (In Russian)
8. Sukhodolov AP (2010) World uranium reserves: prospects for the raw materials supply for nuclear power. Bull ISEA 4(72):166–169 (In Russian)
9. Batenin VM, Zaichenko VM, Leontiev AI, Chernyavsky AA (2017) The concept of the distributed energy development in Russia. Bull Russ Acad Sci: Power Eng 1:3–18 (In Russian)
10. Shapovalov AB (ed) On the issue of carbon-containing fuels. In: 2 parts/under the ed. of Semenov AV, Malyshev NG, Rudenko Yu S (ed) Sustainable development: society, ecology, economy. Materials of the XV international scientific conference. PEIHE "Moscow Witte University", pp 1–748, pp 195–206 (In Russian)
11. Shapovalov AB (2017) Some specificities of power consumption value based on nano-decomposition of carbon-containing waste. In: 2 parts/under the ed. of Semenov AV, Malyshev NG, Rudenko Yu S (ed) Ecological and environmental problems of modern society and ways to solve them. Materials of the XIII international scientific conference, PEIHE "Moscow Witte University", pp 1–504, pp 105–114 (In Russian)
12. Weinstein EF, Shapovalov AB (2010) High-rate pyrolysis as a method for producing nanosized particles. Ecology of industrial manufacturing, vol 3, pp 47–50 (In Russian)

The Criterion for the Existence of Ecology as an Area of a Chain Process of Continuous Directed Self-organization

Aleksandr B. Shapovalov

Abstract The transformation of the primary energy of our planet by the continuous chain process of self-organization of its endogenous substance has formed an area of continuous directed multidirectional self-organization of biota and society. This area as the essence of life unites numerous definitions of the term "ecology". This approach made it possible to formalize ecology as an area of continuous chain self-organizations of biota and society in the form of tensor spaces (fields). The condition for the physical existence of ecology as an area of a chain process of continuous directed self-organization (life) is determined by the criterion that limits the tensor gradients of biota and society by nabla $\nabla \neq 0$.

Keywords Ecology · Biosphere · Biota · Society · Energy · Continuous self-organization · Directed self-organization · Tensor fields · Nabla

The unsustainability of the processes of energy transformation in society is reflected in the dominant concept of sustainable development of the WCED Commission "G. H. Brundtland" since 1987. However, the sustainability imposed on the processes [1, 12] leads to the accumulation and concentration of the conflict potential, and as a result to its catastrophic resolution. Probably, the spread of this very concept marked [1, 12] the today's global socio-economic crisis, which raises doubts about the validity of this approach. This is what actualizes the search for conditions for the transformation of energy that preserves society.

The materiality of nature, confirmed by numerous experiments, is presented as a system of elementary excitations of various fields in the form of particles (quanta), which are the carriers of their superior substances interaction. Thus, the transformation of energy, as the basic axiom of the existence of forms of matter, determines the presence, content and correlation of processes in nature and, therefore, in society.

In the light of current ideas, the primary energy of our planet (Fig. 1), is mainly due to the commuting of the energies absorbed from meteors and meteorites, from

A. B. Shapovalov (✉)
Moscow Witte University, 12, 2nd Kozhukhovsky Pas., Bldg. 1, Moscow 115432, Russia
e-mail: shapovalov-ab@yandex.ru

A. V. Semenov and I. A. Sokolov (eds.), *Sustainable Development: Society, Ecology, Economy*, Earth and Environmental Sciences Library,
https://doi.org/10.1007/978-3-030-73110-6_2

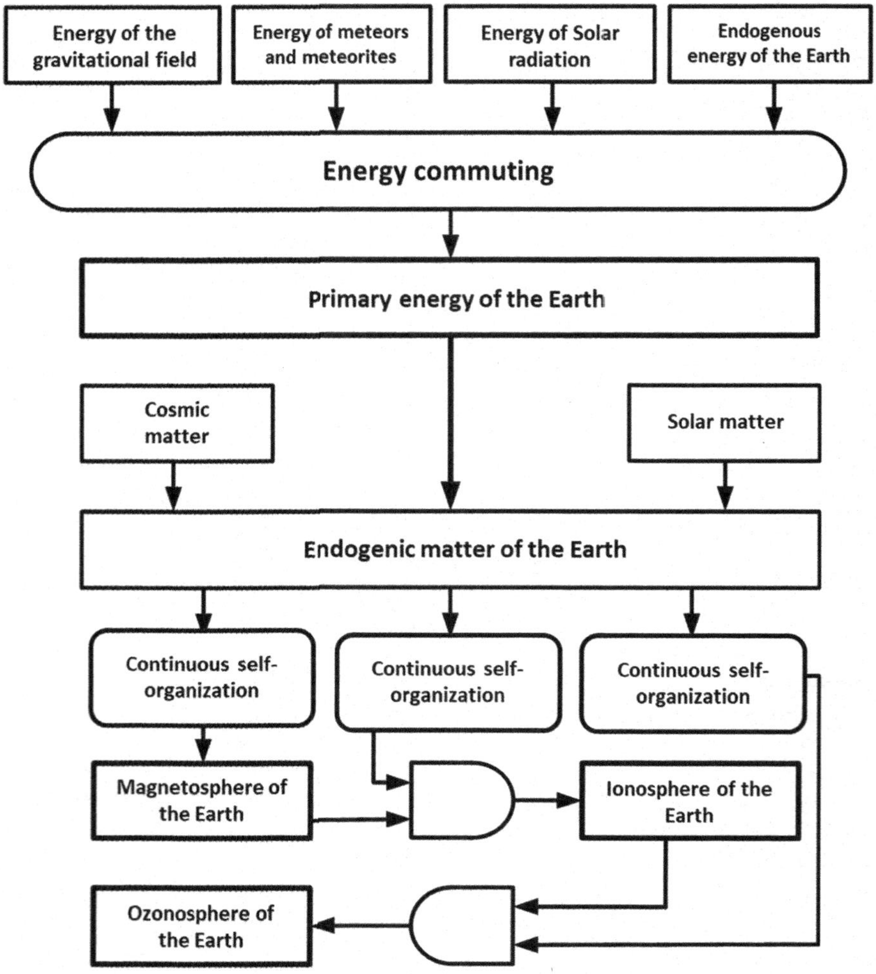

Fig. 1 Continuous chain self-organization of the Earth's geospheres

the gravitational field, from endogenous processes of the Earth and from the radiation of the Sun in correlation with the energy of space objects.

The primary energy of the Earth (Fig. 1) is transformed by its endogenous substance. As a result of chain self-organization, sustainable spatio-temporal forms of matter [2, 3] emerge in the form of spheres and sporadic layers. The emerged spheres adapt to the energy flows by individual continuous self-organization. Thus, the chain self-organization in the atmosphere of dynamically sustainable forms of matter, both endogenous and introduced from outside in the form of sustainable spatio-temporal forms with manifested anisotropy is experimentally confirmed [2, 3]. As a result (Fig. 1) of the continuous chain self-organization of the substance of our planet, the magnetosphere and the elements of the geosphere were formed.

The atmospheric elements of the geosphere form a multilayer inverter energy filter of the Earth, which determines the transformation of the endogenous matter of the planet. Moreover, the geosphere forms a spherical resonator of electromagnetic waves. The detected standing "Schumann waves" at altitudes up to 850 km [10], probably [11], is a repository of biorhythm standards, i.e., their kind of camertone.

Precisely after the formation of the ozone layer (500–600 million years ago), the chain process of self-organization of highly organized and socially structured biological forms from plants to the species Homo sapiens occurred.

The transformation of energy by the endogenous substance of our planet through the breakdown and new chemical bond formation is precisely due to the redistribution of electrons and atomic nuclei. The criterion for this redistribution of chemical elements can be determined by the constancy of their total number and isotopic composition.

The syntheses initiation emerges from CO_2, H_2O, N_2 and mineral salts. The self-organization of *sustainable forms of carbon-containing biopolymers* (nucleic acids, proteins and polysaccharides), *as well as mixed biopolymers* (glycoproteins, lipoproteins, glycolipids, etc.) within the boundaries that ensure their physical existence and determines (Fig. 2) the *biosphere.*

Adaptability and mutual correlation "sociology" [4] as forms of self-organization of biopolymer systems to a greater extent than covalent ones determine less strong bonds (hydrogen bonds are by 15–20 times less strong than covalent ones). Therefore, it is these bonds that realize the self-organization of sustainable spatio-temporal biopolymer systems in relation to energy flows. Moreover, therefore, precisely these bonds that provide the mutual chain [4] self-organization of spatio-temporal biopolymer systems in the form of their mutual correlation and hierarchical structuring.

Therefore, from the whole variety of transformation of energy and matter, the only form of energy distribution in the biosphere is different from the cycle, in the form of [5, 6] unidirectional flow. The continuity of this flow *of self-organization of matter* in the biosphere is obvious and represents *life.*

Precisely the transformation of energy through chemical bonds with the formation of *sustainable spatio-temporal forms of carbon-containing biopolymers* within the boundaries that ensure their physical existence, determined the *biosphere.*

Continuous chain structural self-organization of biopolymer forms leads to the formation of sustainable spatio-temporal living systems and their hierarchical and correlation (socialization) self-organization, which ultimately form the biomass of the biosphere.

The continuous flow of energy through the chemical bonds of biomass maintains a sustainable imbalance (homeostasis) between living organisms and the environment (biogeocenoses), through biogeochemical cycles and the structure of the biosphere. The dynamics of biomass is determined by biocycles of producers, consumers and decomposers, and by long-term abiogenic and biogenic (metamorphism) processes.

The energy accumulated by the chemical bonds of biomass, by its natural oxidation is partially released and diffused (dissipation) into outer space (several billion so-called oil equivalents, annually), and partially accumulated. Abiogenic and biogenic

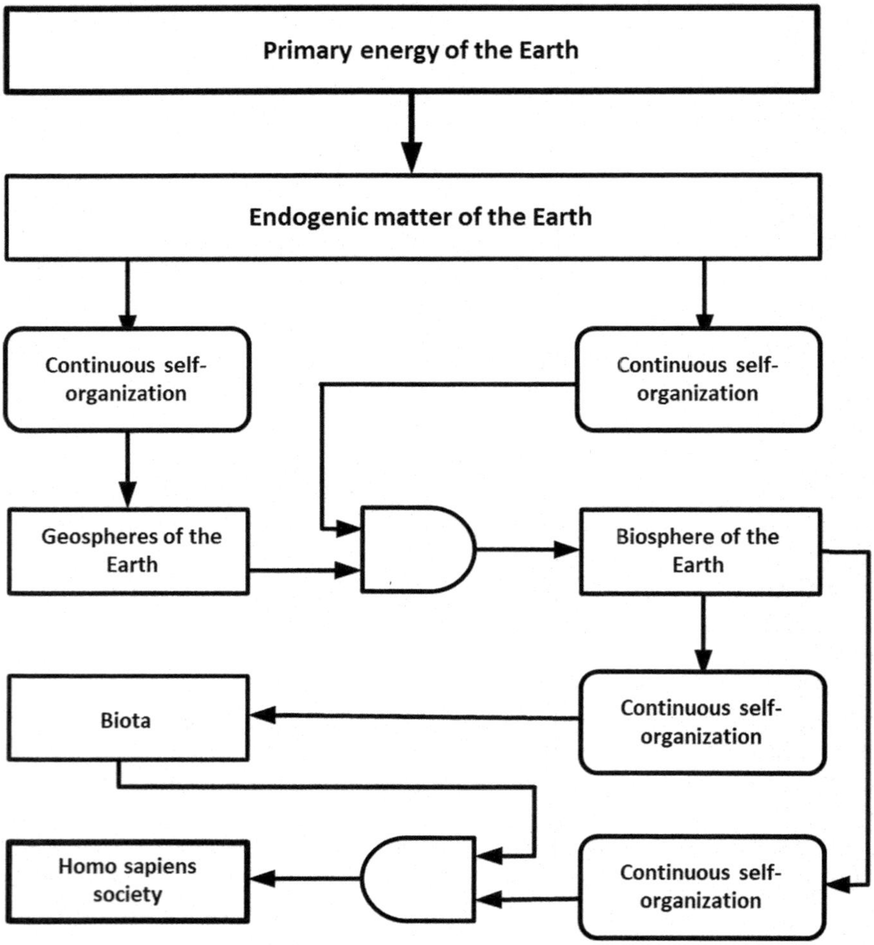

Fig. 2 Self-organization of endogenous matter in the biosphere of the Earth

(metamorphism) processes form carbonaceous energy carriers (CEC) accumulated in the biosphere in the form of various, including fossil, energy carriers (peat, oil, gas, etc.).

Continuous chain self-organization sublimated the biomass with the appearance of Homo sapiens species (Fig. 2). Socialization, as a concentrator of energy, Homo sapiens within the framework of society made it possible, on the basis of the production, accumulation, processing and synthesis of information [7], to purposefully affect the transformation of energy and matter in the biosphere (anthropogenic impact). The involvement of the resources of the biosphere and on this information base, and first of all, of the accumulated CEC intensifies continuous self-organization and, therefore, determines the state and dynamics of the development of society.

The self-organization of Homo sapiens biological species by socialization singled out its existence in the biosphere into an artificial environment created by it. Moreover, the state and dynamics of processes in society is determined by its specifics: information flows, dynamics of ideologemes, ways of their implementation, etc., which is completely absent in (Fig. 2) the rest of the biosphere—biota. Therefore, it is logical to assume that the emerged society is a *separate* open nonequilibrium dynamic unidirectional chain process of continuous self-organization of carbon-containing forms of matter with its *own direction of continuous self-organization.*

Thus, the modern transformation of energy and matter in the biosphere is realized by separate processes with *various directions of continuous self-organization.*

The total [8] biomass of all living organisms of the biosphere is currently estimated at about 550 billion tons of carbon. The total biomass of Homo sapiens species is estimated to be only 0.06 billion tons of carbon, which is approximately 0.01% of the total biomass of the biosphere. Masses, as indicators of inertia, determine the dynamics of energy transformation processes. It is easy to see that the inertia of the biosphere process is by more than 9000 times greater than the inertia of the process of society.

Fluctuations in the direction of continuous self-organization in society, often in the form of generated dubious ideologemes and goals that determine even more dubious ways to achieve them, are counterproductive and dangerous, and not only waste huge resources, but are also catastrophic both for society and the biosphere as a whole. This actualizes the definition of the boundaries of the existence of society and biota in the biosphere.

Continuous directed self-organization, defining the essence of life, unites numerous contents of the term "ecology" since (E. Haeckel) 1866. Based on this, it is logical to define: *"ecology is the area of continuous chain directed self-organization of biota and society."*

The formalization of real ecology makes it possible to reveal consequences from a finite set of axioms (empirical laws, hypotheses, idealizations, simplifications, etc.) by deduction in the form of forecasts, informational and quantitative assessments.

The space [9] defined by a set of objects in the form of elements or points (functions, geometric shapes, physical states, etc.), with relations, which are introduced between them axiomatically, can determine the geometry of the spaces of biota and society.

Based on this, let us represent biota and society in the form of nonvacuous sets, where each pair of their elements is associated with a non-negative real number that satisfies a finite set of axioms:

B—space (field) of biota,

S—space (field) of society,

E—space (field) of ecology.

The properties of spaces (biofields) **B** and **S** are expressed by gradients that physically indicate the direction of their greatest growth, regardless of coordinates (invariance). The anisotropy, parameter set and correlation of the parameters of the multidimensional spaces **B** and **S** determines the array of parameters (components) together with the laws of transformation of parameters (components) when the coordinates (basis) are varied, which is modeled by the tensor at each point of the considered space. Imaging, which assigns a tensor to each point of the space under consideration, forms the tensor field of the spaces B and **S**.

Based on this, the properties of the spaces B and S, represented in the form of tensor fields, are expressed by the corresponding [9] tensor gradients - the nabla ∇. This representation makes it possible to determine the *continuity of directed self-organization of spaces* **B** and **S**, as a condition limiting their tensor gradients by an operator, nabla $\nabla \neq 0$.

Moreover [8], the nabla affects on the fields to the right of it, and does not affect on the fields to the left of it. Therefore, [9], the scalar and vector products involving nabla are not commutative, and it does not belong to the space of ordinary vectors. Thus, nabla is not charactsrized by the properties of ordinary vectors and, therefore, does not correspond to the geometric properties of an ordinary vector, in particular, it does not commute with vectors. In relation to biofields, this can be interpreted as the unidirectionality of the life flow (including the genetic code) of spaces **B** and **S**.

This makes it possible to formalize ecology as an "*area of continuous directed self-organization of biota and society*" in the form of:

$$E = B \cap \mathrm{S} \forall \nabla \neq 0.$$

The proposed formalization will determine the condition for the physical existence of life in the field of chain continuous directional self-organizations of biota and society in the form of a specific criterion. Moreover, such a formalization allows experimenting with the spaces E, B and S, which is impossible in reality and investigate responses to changes in parameters and initial conditions. Compliance with this criterion for any variation is a kind of "*ecopolis*" of preserving the life of the biota and society, as well as the biosphere as a whole.

The work was done on our own initiative and at our own expense.

References

1. Snakin VV (2016) The path to sustainable development: myth and reality. Age of Globalization. MSU, № 1–2, pp 83–86
2. Zelensky LM, Veselovsky IS, Fizmatlit M (eds) (2008) Plasma heliogeophysics, vol 2, 560 pp
3. Miroshnichenko LI (2011) Physics of the Sun and solar-terrestrial relations. In: Panasyuk MI (eds) Universitetskaya Kniga (University book), 174 pp
4. Lehn JM (1998) Supramolecular chemistry: concepts and perspectives. Nauka, Novosibirsk, 334 pp
5. Vernadsky VI (1998) Philosophical thoughts of a naturalist. Nauka, 520 pp

6. Pechurkin NS (1988) Energy and life. Nauka, Novosibirsk, Sib. department, 190 pp
7. Shapovalov AB (2019) On the issue of carbon-containing energy carriers. In: 2 parts Semenov AV, Malyshev NG, YuS (eds) Sustainable development: society, ecology, economics. Materials of the XV international scientific conference; "Moscow Witte University". Part 1, 748 pp, pp 195–206
8. Bar-On YM, Phillips R., Milo R (2018) The biomass distribution on Earth. Proc Natl Acad Sci 115(25):6506–6511
9. Korn GA, Korn ThM (1968) Mathematical handbook for scientists and engineers: definitions, theorems, and formulas for reference and review. McGraw-Hill Book Company, New York etc., pp 1130
10. Simões F, Pfaff R, Freudenreich H (2011) Satellite observations of Schumann resonances in the Earth's ionosphere. Geophys Res Lett 38(22)
11. Ludwig W (2010) Informative Medizin. Verlag für Ganzheitsmedizin, Essen, p 240
12. Shapovalov AB, Smurov AV (2020) Ecology and nanotechnology. Sci Pract J "The Life of the Earth", vol 42, № 1. University Press; MAX Press, Moscow, pp 38–45

Environmental Factor in Assessing the Investment Attractiveness of the Region

Alexandra V. Voronina

Abstract The attractiveness of the Rostov region from the point of view of safe investment in the regional economy within the framework of ecology is considered in the paper. The general and environmental factors affecting the investment potential and attractiveness of the region are reviewed. The two types of factors affecting the natural and environmental component are highlighted among the others. These are natural resource and environmental factors. The environmental policy of the region in the field of expanding the areas of attracting investors to the Rostov region is considered. Air pollution is considered a major environmental problem in the region. Sources of pollution are vehicles and energy facilities. The amount and sources of funding for the elimination of pollution factors are investigated in the work. The criteria and indicators for evaluating investment attractiveness are characterized, taking into account the environmental factor. The method of calculating of the integral and environmental coefficients of investment attractiveness is given.

Keywords Investment · Investment attractiveness · Environmental factor · Coefficient of environmental attractiveness

1 Introduction

The attractiveness of the region from the point of view of potential investors has a significant impact on the level of socio-economic development of the constituent entity of the Russian Federation. Various events are implemented, participation in exhibitions, various forums, meetings and conferences are held to attract investments. Regional laws and regulations are adopted, competitive advantages are formed, which significantly improves the investment climate in the region, develops investment potential and affects the reduction of investment risks.

A. V. Voronina (✉)
Moscow Witte University, branch in Rostov-on-don, 78 Nefedova str., Rostov-on-don 344013, Russia
e-mail: caas@mail.ru

A. V. Semenov and I. A. Sokolov (eds.), *Sustainable Development: Society, Ecology, Economy*, Earth and Environmental Sciences Library,
https://doi.org/10.1007/978-3-030-73110-6_3

The purpose is to study the level of investment attractiveness of the Rostov region and the impact of environmental factors on it.

Empirical and hermeneutical analysis, statistical review, the method of analogies were used as the investigation methods. Various methods of cognition, interpretation, systematization and generalization of information are used during the research.

The main results of the study can be used as a basis for further thorough research of investment of environmental risks and investment of environmental attractiveness.

2 Factors of Investment Attractiveness

The main scientific issue in this context is to identify the factors that affect the investment attractiveness of the constituent entity of the Russian Federation. The summary of scientific works of Russian and foreign authors on the subject of systematization of these factors allows us to draw a conclusion about the existence of a significant number of different typologies. Typologies and classifications presented in the scientific works sometimes correlate with each other, sometimes drastically differ, and initially it is due to the interpretation of the definition of "investment attractiveness" and its derivatives, such as "investment climate", "investment activity", "investment risks", etc.

Most of the authors agree that to assess the investment climate and investment attractiveness, a factor approach is necessary, which allows us to take into account the specificities of the regional environmental climate, the demographic component and the features of the regional economy. In addition, it seems that the use and operation of statistical data reduces the degree of subjectivity in prognosing the development of investment attractiveness. It should be noted that the investment attractiveness of the region includes the competitiveness of each municipality included in the constituent entity of the Russian Federation. The investment attractiveness of a municipality should also be considered from the point of view of investment potential and environmental risk [10].

Summarizing the above, we can h the factors presented in Fig. 1.

Among the two groups of factors, there are two types that affect the natural and environmental component. These are natural resource and environmental factors.

They are as important as other factors for the Rostov oblast, and in some cases, they are even key factors. For example, in cases when making decisions about investing in the agricultural industry. Rostov region is located in the zone of moderate continental climate with an average annual precipitation of 400-650 mm per year. Average temperatures in January vary from −9 to −5 °C, and in July from +22 to +24 °C. The region is situated in the steppe zone and it is rich in natural land, water and mineral resources. The territory of the Rostov region is rich in warmth and sunlight—the duration of the period with a temperature above 10 °C lasts from 160 to 180 days. The main port facilities of the Rostov region are located in Rostov-on-don, Taganrog,

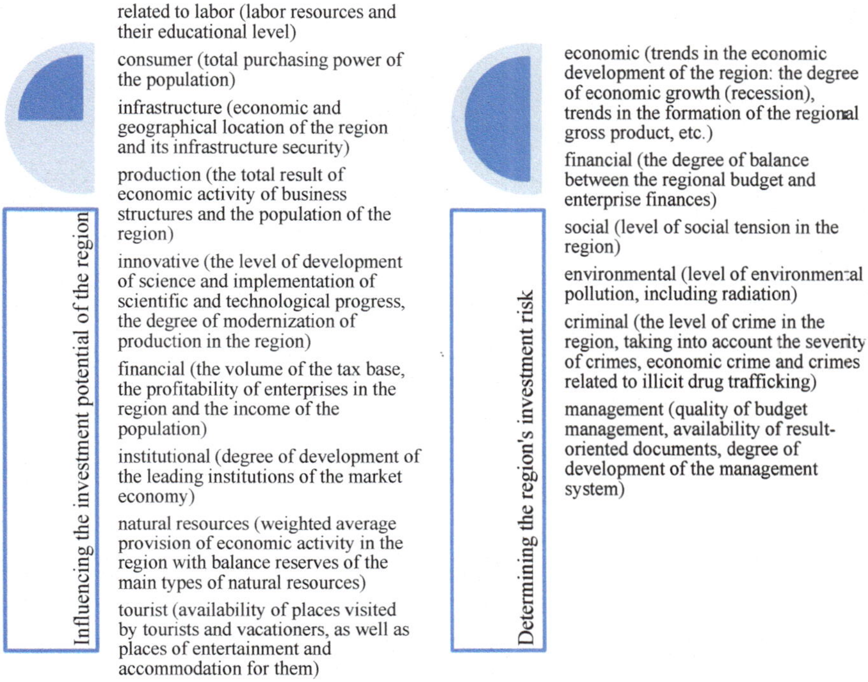

Fig. 1 Factors affecting the investment attractiveness of the region

Azov, Volgodonsk, and Ust-Donetsk. Taganrog, Rostov and Azov ports are international and connect the region with five seas, providing access to the countries of the Mediterranean and Western Europe [5].

3 Environmental Policy of the Region

Rostov region is one of the most industrially developed regions of Russia where the country's largest industrial enterprises are located: metallurgical, machine-building, and energy. Economic success, as everywhere in the world, entails a number of environmental problems. This includes excessive use of natural resources, pollution of the biosphere, and the problem of waste.

Air pollution is considered to be a major environmental problem in the region. Sources of pollution are vehicles and energy facilities. The combustion of fuel sources entails the emission of harmful substances into the atmosphere. Despite the fact that enterprises use waste treatment plants, polluting particles still get into the environment.

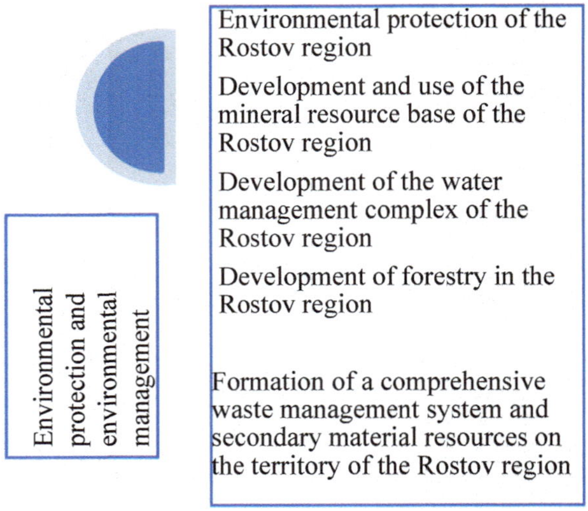

Fig. 2 Directions of the Environmental Protection Program

Waste and garbage, sources of air, water and soil pollution are of no less danger. The region has a large number of landfills, however, their content does not meet sanitary standards. Quite often there is a fire of waste due to their congestion, and chemicals are released into the atmosphere. Unfortunately, there are only three enterprises in the region that sort trash. In the future, the raw material can be recycled.

Until 2017, the Rostov region was in the group of regions with medium investment potential and moderate investment risk. Thus, the environmental factor in the region increases the investment risks of the Rostov region.

In order to minimize investment environmental risks, the government of the Rostov region has been taking a number of measures for several years. The environmental policy can be based on the Strategy for Preserving the Environment and Natural Resources of the Rostov Region untill 2030 [1] and the state Program of the Rostov region "Environmental Protection and Rational Use of Natural Resources". The program is implemented in five directions, shown in Fig. 2 [2].

The following actions have been taken by the Government of the Rostov region within the framework of the implementation of measures aimed at reducing environmental investment risk [6]:

(a) more than 30 km of riverbeds have been cleared, and the estimated prevented damage is 961 million rubles.;
(b) five hydraulic structures in four municipalities of the region were repaired, the estimated prevented damage—135 million rubles.;
(c) educational activities are carried out to increase the environmental awareness of students;
(d) more than two thousand kilometers of water protection zones borders and coastal shelter belts have been established, as well as 240 special information landmarks have been installed.

Table 1 Current (operational) costs of environmental protection, thsd. rubles

Expenditure areas	2017	2016	2015	2014	2013
The costs of environmental protection in total	6178799	5941235	5364009	5329572	5106197
Including: protection of atmospheric air	320356	321493	303027	309985	333061
Waste water collection and treatment	3856031	3648076	3304053	3191975	3029817
The protection and rehabilitation of land, surface and groundwater	54960	44426	56894	56247	64255
Waste management	1661569	1590302	1440712	1440718	1370593
Protection of the environment from various types of physical impact	2641	1921	2645	2455	1906
Conservation of biodiversity and protection of natural territories	8394	5535	5383	810	284
Research and development activities to reduce negative impacts on the environment	128	20	4882	9877	3952
Ensurance of radiation safety	247002	291590	208785	234746	223239
Other areas in the field of environmental protection	27718	37872	37628	82759	79090

Due to the constant measures undertaken to reduce the environmental risk, there is an increase in investment in the region. This fact is confirmed by statistical data [9].

The costs of environmental measures for five years are grouped in Table 1 [11].

Data from Table 1 show that the total cost of environmental protection measures increased by 1073 thsd. rubles during the study period. A significant increase is observed in areas related to the collection and treatment of waste water and waste management.

The installation of an automatic station for monitoring atmospheric air pollution in real time near the Rostov-arena stadium may serve as an example of measures undertaken for the protection of atmospheric air.

Among the ways to reduce the ecological risk are the publication of Don Ecological Bulletin, the conduction of more than two thousand environmental annual clean-ups and implementation of more than five thousand environmental education programs with the purpose to increase the ecological culture of the population.

Among the successfully implemented investment projects, taking into account environmental factors are the following: organization of production of high endurance polyolefin film threads in Kamensk-Shakhtinsky, organization of production of snacks from fish and squid, with a volume of 3600 tons of end products per year in Novoshakhtinsk, reconstruction and modernization of poultry farms with a capacity increase up to 145 mln eggs, 500 thsd. chickens, hens, two thousand tons of meat per year in Belaya Kalitva, reconstruction of an irrigated area of 2200 hectares in Bagaevskaya region, construction of an electric desalting plant and atmospheric vacuum distillation of oil in Krasniy Sulin district, etc.

4 Indicators and Indices for Assessing Investment Attractiveness Taking into Account the Environmental Factor

The gross regional product can be used as an integral criterion for the level of investment attractiveness of the region. Its use can be justified by the fact that the gross regional product is a summarizing indicator of the development of the region's economy. Information about it is available in the public domain. It is homogeneous by industry structure and indirectly takes into account the key indicators of the current and future competitiveness of the region.

Theoretically, in conditions of equal provision of regions with the basic factors of investment attractiveness, the higher the degree of the efficiency of the basic factors use, the more competitive the region is (a delta of competitiveness is formed between more and less competitive regions, corresponding to the difference in the level of efficiency of resource use). At the same time, the prevalence of one or another factor in the economy indicates the level/stage of economic development [3].

The attractiveness of the region at the level of innovation-oriented industries is determined by the ability to introduce innovative technologies and use them to produce goods with unique consumer properties.

The integral indicator of investment attractiveness is calculated using the multivariate mean formula (1):

$$Q_i = \frac{\sum_{s=1}^{C} K_s X_{si}}{\sum_{s=1}^{C} K_s} \tag{1}$$

where Q_i is the integral level of investment attractiveness of the i—industry, comparable to the average level for the industry;

i-1 … n—number of main industries;

n—number of industries;

s-1 … C—reducible specific indicators;

C is the number of reduced indicators;

K_s—weight number (score) of the s-th indicator;

X_{si}-numeric value of the s-th indicator for the i-th industry.

The integral level of investment potential is determined by combining individual indicators and then dividing them in the number of values that are taken into account [4].

The environmental attractiveness of the region can be determined using the environmental attractiveness indicator, which is the ratio of the fertilizers amount (mineral and organic) applied per 1 ha to the standard indicator and taking into account the

Table 2 Main indicators characterizing investments in the first quarter of 2019, mln. rub

Indicators	Amount of investments	2019 in % to 2018	2018 in % to 2017
Financial investments, total	129906	155.2	133.1
including: long term ones	5308	91.0	160.4
short term ones	124598	160.1	131.4
Investments in fixed assets	29511		
Investment volume index		87.4	83.3

anthropogenic burden on agricultural areas of the Rostov region, depending on the division into impact, transition and background zones [7] according to the formula (2):

$$K_{ia} = K_{mf} + K_{of} + K_{ab} \tag{2}$$

where K_{mf} is the coefficient calculated as the ratio of the amount of mineral fertilizers applied per 1 ha to the standard indicator;

K_{of}—coefficient calculated as the ratio of the amount of organic fertilizers applied per 1 ha to the standard indicator;

K_{ab}—coefficient of anthropogenic load, calculated using the formula (3):

$$K_{ab} = 0,2S_{iz} + 0,5S_{tz} + S_{bz} \tag{3}$$

where S_{iz}—is the area of the regions included into the impact zone, a share of the area of the region;

S_{tz}—the area of the regions included into the transition zone, the share of the area of the region;

S_{bz}—area of the regions included into the background zone, share of the area of the region;

0,2 and 0,5—are weight numbers determined by expert analysis.

According to the results of the study by the National Rating Agency, the investment rating of the Rostov oblast has not changed over the past two years. The oblast is included in the category with average investment attractiveness and low investment risk.

For the first quarter of 2019, the Rostov region reached the investment indicators shown in Table 2.

Data from Table 2 indicate an increase in investment activity in the Rostov oblast, and first of all, this is indicated by the investment volume index, which has increased on 4.1% over the past two years.

It should be noted that the growth of investment attractiveness of the industry and the economy of the region as a whole should be accompanied by significant investments, minimizing risk, including environmental risk.

5 Conclusions

As a part of the environmental factor impact in assessing the investment attractiveness of the region, the Ministry of Natural Resources and Ecology of the Rostov oblast is monitoring four investment projects, three of them established cooperation agreements with the participation of the government of the Rostov oblast following the results of the II international exhibition-forum "ECOTECH". The total volume of investments in environmental projects in 2018 amounted to 1245 mln. rub.

Thus, from the point of view of the environmental factor impact, the competitive advantages of the Rostov region in order to attract investment are:

- advantageous geographical position;
- geopolitical importance;
- high natural resource potential and a rich mineral resource base.

Preservation of the environment, its restoration and improvement, environmental education of young people is also the patriotic and cultural education. Therefore, ecological culture is a necessary and very important component in the structure of human activity [8]. Underestimating the environmental and natural factors leads to a decrease in the investment attractiveness of the region. This fact should be taken into account when building strategic plans for the development of the region. Indicators of the Rostov oblast in this direction are steadily increasing, as evidenced by a change in its rating in 2017 from the region with medium risk investment to the region with a low risk investment and a steady growth of investments.

References

1. Strategy of Investment Development of the Rostov Region until 2030: Resolution of the Government of the Rostov Oblast from 31 July 2013 № 474 (in the edition of the annex to the Government Decree from 28 March 2018 № 197). – [Electronic resource]. http://old.donland.ru/Donland/Pages/View.aspx?pageid=75189&mid=128186&itemId=119. Accessed 03 May 2019
2. On the Approval of the State Program of the Rostov Oblast "Environmental Protection and Rational Use of Natural Resources": Resolution of the Government of the Rostov Region from 15 October 2018 № 638 (in the edition of the annex to the Government Decree from 17 May 2019 № 346). – [Electronic resource]. – Mode of access. http://docs.cntd.ru/document/550217692. Accessed 05 June 2019

3. Voronina AV, Goncharova SN (2016) Indicators for assessing regional competitiveness. Transport: Science, Education, Production: Proceedings of the International Scientific and Practical Conference, pp 63–67
4. Davydova LV, Ilminsky SA (2013) Evaluation of investment attractiveness of the region. Finance and Credit. №11(539). [Electronic resource]. https://cyberleninka.ru/article/n/otsenka-investitsionnoy-privlekatelnosti-regiona-5. Accessed 03 April 2019
5. Ivashchenko KA (2019) Investments in the Rostov region. In: Materials of the XI International student scientific conference “Students’ Scientific Forum”. – [Electronic resource]. Mode of access: https://scienceforum.ru/2019/article/2018015842. Accessed 03 July 2019
6. Lysochenko AA (2017) Ensuring environmental safety of the Rostov Oblast as a factor of strategic development of the region. Reg Econ Manag: Electron Sci J № 2(50). [Electronic resource]. Mode of access: https://eee-region.ru/article/5017. Accessed 17 May 2019
7. Moskalenko AP, Moskalenko SA, Gubachev VA (2013) Ecological factor in assessing the investment attractiveness of agricultural nature management as an economic form of organizational support. Manag Econ Syst: Sci Electron J 2(50):39–52
8. Oganesyan SV, Alekseeva SA (2012) The role of higher education in the forming of cultural and moral priorities and environmental awareness of students. Bull Moscow Witte Univ Series 1: Econ Manag 2:105–109
9. Abazieva KG, Goncharova SN, Stratan DI, Voronina AV, Shevchenko MV (2017) Strategy of import substitution in the Rostov region: results, problems, prospects. Int J Appl Bus Econ Res 15(23):495–509
10. Voronina AV, Osipyan NB, Dmitrieva MA, Elchaninova OV, Vatolina MV (eds) Legal and economic methods as an environmental risk management mechanism. Res J Pharmaceut Biol Chem Sci 9(6):1671–1677
11. Website of the Territorial Body of the Federal State Statistics Service for the Rostov Oblast. [Electronic resource]. Mode of access: http://rostov.gks.ru/. Accessed 26 April 2019

Crucial Issues of Management of Social and Economic Systems in the Aspect of Sustainable Development Concept

Sustainable Development: The Definition of the Essence and Quality Integrity

Boris V. Salikhov

Abstract The purpose of the article is to identify the essence of the phenomenon of sustainable and unsustainable development. The relevance of the topic is due to the dysfunctions of the current scientific and practical paradigm of the problem. The methodological and methodical concept of the study is characterized by both a systematic and logical-epistemological approach to the definition of the essence of sustainable development, as well as the unity of qualitative and quantitative analysis. The main scientific results consist in the definition of the essence and relevant forms of sustainable and unsustainable development; in the scientific assumption that the modern Russian civilization is in a state of unsustainable regressive development, which determines the imperative of radical institutional innovations.

Keywords Sustainable development · Unsustainable development · Sustainable progressive development · Unsustainable regressive development · Qualitative integrity

1 Introduction

The issue of sustainable development has long gone beyond academic discourse and has acquired practical forms. This is due to the growing number of interrelated global, national and regional problems. In this regard, various international and regional structures, especially the UN, actualized the mentioned problems by developing a system of goals, objectives and indicators of sustainable development of society, economy and nature. Meanwhile, despite the efforts, systemic problems of social and economic development continue to grow in the world [3, 5, 8]. It is reasonable to assume, that existing research paradigm does not allow achieving the

---

B. V. Salikhov (✉)
Moscow Witte University, 12, 2nd Kozhukhovsky pas., Bldg. 1, Moscow 115432, Russia
e-mail: mgsusalikhov@yandex.ru

Moscow State Linguistic University, 38, Ostozhenka st., Bldg. 1, Moscow 115432, Russia

A. V. Semenov and I. A. Sokolov (eds.), *Sustainable Development: Society, Ecology, Economy*, Earth and Environmental Sciences Library,
https://doi.org/10.1007/978-3-030-73110-6_4

sustainable development goals. First of all, alarming the utilitarian and monodisciplinary interpretation of the sustainable development essence. In the first case, we are talking about maintaining the level of consumption of present and future generations of the planet's inhabitants (UN option). In the second aspect, sustainable development is linked to the sustainable economy development and the preservation of natural capital [1, 13].

In addition, there is an obvious incompleteness of the study, where the person "disappeared", which contradicts the imperative of inclusiveness, that is, the scientific and practical requirement "to reach every person" [7, 11]. Seventeen goals that set out in UN documents and have received further interpretation in the relevant international documents are not focused to a certain results that ensure the sustainability of human life. But most importantly, the current paradigm of sustainable development research does not contain an endogenous mechanism of interest of key actors in the high efficiency of solving the relevant problems [2, 10]. Hence, the subject of this article is to identify the essence of sustainable (unsustainable) development and, consequently, its main relevant modified forms. This will make it possible to clearly define and adjust the scientific and practical perspective of interdisciplinary understanding and systematic solution of applied problems in the field of sustainable development as such.

2 Methodology and Logic of Research

The methodological approach to identifying the essence of sustainable development is the epistemological analysis of the "sustainable development" phenomenon. In the framework of this approach, separately, *sustainability and development* in their subsequent integration. "Essence matters", and it is not so much academic, but practical problem, because it is always "deployed" in specific content elements. And the more these elements are revealed, the more complete the result of the study of the problem [9]. Here, the dialectics of *the phenomenon* and *the essence* is seen as a critically important method of research. Only in this case there is a chance to avoid the temptation to consider "visibility" instead of the real "phenomenon" [4]. Unfortunately, in the term of sustainable development part of the discourses is in the "visibility" and not "phenomenon", that determines the overall low research performance.

In addition, the dialectical method aims to find the essence of the opposite processes, in particular, the phenomenon of *unsustainable development.* This will provide the right perspective for analytical thinking, as sustainable development can be seen as a function of activities in the way of overcoming key elements of instability. In this case the relevant scientific and practical activity of the researcher acquires the necessary specifics. Further, sustainable and unsustainable development may have different shapes, which directs the researcher to identify these features in order to identify the stages of ascent to the parameters and indicators of sustainability. It is reasonable to note that the development can be both on an upward trajectory and a downward vector.

Since even within the framework of the current paradigm the concept of sustainable development includes nature, economy and society, it is obvious that all elements of the methodology of the system paradigm are included in the analysis [9]. In other words, the interpretation of the essence of this concept should include various elements that have "external integrity and internal diversity" [9, P. 109]. This means that the sustainable or unsustainable development is always a certain set of internal and external conditions and factors. Consequently, the problem of sustainability or unsustainability of development is the problem of identifying, clearly defining and skillfully combining the specific conditions and factors that ensure the effectiveness of relevant decisions.

It is important to emphasize, that conditions and factors of sustainable development should not be monodisciplinary. Even within the framework of the current UN "triad" (society, ecology, economy), an interdisciplinary discourse is needed, including the analysis of historical, mental, socio-cultural, institutional and other conditions and factors [1, 14]. This means that sustainable development always corresponds to a certain *qualitative integrity* of the relevant conditions and factors that are in a state of controlled and dosed interaction. Otherwise, objectively, the development can be neither sustainable nor progressive. It can be added that the qualitative integrity of the conditions and factors of sustainable development should correspond to their certain *qualitative quantity* as a specific set of these conditions and factors "here and now" [1, 15].

3 Discussion of the Problem and the Results

There are different approaches to the interpretation of the essence of sustainable development. In the literature, sustainability is mainly correlated with economic growth and also interpreted as not a decrease in the total capital of the nation over time [1, 6]. However, dysfunctions of the monodisciplinary approach aim at a more in-depth understanding of the problem, including many processes [2,5,10,]. It is important to note that the UN paradigm is too abstract and the scientific discourse on sustainable development should be added at least in two aspects. First, we need to expand the scope of qualitative quantification of conditions and factors that determine the nature of sustainable development. Secondly, it is necessary to fix the progressive target function of this development. Such function can be formation and dynamics of *worthy life of the person.* If a worthy life is understood as a moral form of the standard of living, then a clear picture of the interdisciplinary perspective of the problem is formed [8, 11, 14].

Now it should be outlined the epistemological framework of the category "sustainable development". First of all, the *sustainability* is always understood as the ability of the system to maintain its state or quickly return to the previous equilibrium position after the deviation. In turn, the *development* is understood not only as "essential, necessary movement, change of something in time", but also as "stable existence on its own basis" [12, P. 463-464]. It is important to note that development can be ascending and descending with the key role of *endogenous conditions and factors.* These provisions allow us to make a number of important conclusions concerning the common understanding and diversity of forms of sustainable development.

Firstly, any system (person, household, organization, society, economy, state, etc.) can develop steadily both in ascending and descending trajectories. In the first case, there will be steady growth; in the second case, there will be a steady decline. It is important to note that the steady growth of the system or its steady decline does not mean the linear character of the dynamics. Non-linear trend and oscillatory characteristics of the process can also take place here. In other words, some volatility of the parameters and indicators of the system in this case is such that all the "points" of the general trend are not characterized by back-and-forth, turbulent dynamics.

Secondly, the ontology of sustainable development always consists in the *effectiveness of endogenous conditions and factors* that determine the internal logic of its dynamics. Any sustainable development is always a function of the endogenous creative potential contained in the internal mechanisms of self-improvement of the system. This means, for example, that ideological choices in the area of any social, institutional, economic, environmental and other transformations should not go beyond the endogenous capabilities of the system. Domestic practice shows a large-scale violation of the principle of endogeneity. It is known that the ideological choice in favor of liberalism and monetarism, which did not correspond to the institutional history of the Russian nation, naturally led to the destruction of our society and the economy [2, 3, 7].

Thirdly, the sustainable development of the system always appears as a *natural result of the interaction of all the actual conditions and factors.* Such interaction should be relevant to the requirements of sustainable development of the system. In this connection, sustainable is always a function of the balance of the relevant conditions and factors. The following forms of interaction may take place: between endogenous conditions and factors; endogenous and exogenous conditions; internal and external factors; exogenous conditions and factors of system development and other. It is obvious that ensuring the sustainable development requires the relevant quality of the entity, that should be able to carry out the selection of the necessary conditions and factors, as well as to organize certain forms of their use [2, 8, 11].

Thus, the essence of sustainable development of any system can be interpreted as its development on the basis of dynamic balance of certain conditions and factors. Dynamic balance of conditions and factors means their synchronicity and complementarity, providing linear (full synchronicity) or nonlinear (incomplete synchronicity) sustainable dynamics of the system. It is obvious that sustainable development can be progressive, neutral and regressive. *Sustainable progressive development* of the system is a development, carried out on the basis of dynamic balance of synchronously improving conditions and factors. *Sustainable neutral development* is based on the simultaneous interaction of conditions and factors that maintain their qualitative integrity and qualitative quantitativeness. Finally, *sustainable regressive development* is an anti-progressive development based on a dynamic balance of conditions and factors in a state of destruction and dysfunction.

Unsustainable development can also be progressive, neutral and regressive. *Unsustainable progressive development* is a turbulent development, based on dynamic disequilibrium conditions and factors that are in a state of progressive but asynchronous improvement. Such development can be in the conditions of radical social, cognitive, technological and other shifts. *Unstable neutral development* can be interpreted as development on the basis of asynchronous interaction of conditions and factors that keep their parameters and indicators volatility unchanged. *Unstable regressive development* can be interpreted as a turbulent development of the system on the basis of dynamic disequilibrium of conditions and factors in a state of continuous and asynchronous degradation. Often the complexity of this situation is to get into a kind of "trap of instability», when the applied control mechanisms, inadequate to civilizational quality of system, form mental model of impossibility to overcome system destructions. An example is the "trap of age-old stagnation" of the world economy [3, 8]. Interpretations of the essence and modified forms of sustainable and unstable development (without neutral forms) are systematized in Table 1.

4 Conclusion

Firstly, the concept of sustainable development is characterized by a plurality of relevant conditions and factors, among which the key are endogenous determinants that ensure the development of the system on its own basis. In practical terms, this means that the subject of management should invest resources, first of all, in the actualization of the internal conditions and factors of development. This requires the most complete consideration of the specifics of the system. When it comes to sustainable development the selection of appropriate conditions and factors should be carried out exclusively within the historical, mental, institutional and other characteristics of a particular civilization. External conditions and factors can give the expected high result only if they correspond to the logic and integrity of internal (ontological) conditions and factors. In particular, liberal ideology in the form of monetarism "does not work" on "Russian soil" precisely because of the natural historical contradictions of Western and Russian civilization.

Secondly, specific efforts to achieve sustainable development will be scholastic if the "matrix of forms" of development is not applied (Table 1). In this regard, the current state of Russian society and economy can be characterized as *unstable regressive development.* This is due to the many back-and-forth of processes, among which the growing public pessimism based on the general uncertainty of people in their future. In addition, there is a turbulent development of the economy (failures in GDP dynamics in 2009 and 2015), demographic and infrastructure problems are growing. At the same time, innovation and investment activity is falling, the middle class is declining, the "poverty trap" is growing, corruption and injustice

Table 1 Interpretations of the essence and modified forms of sustainable and unstable development (without neutral forms)

The nature and form of development	Interpretation of the concept	Features of interaction of conditions and factors
Sustainable development (essence)	Development of the system, based on dynamic balance of conditions and factors	Conditions and factors are, at the same time, in a state of dynamic balance and change
Sustainable progressive development (form 1)	Progressive and creative development of the system, based on dynamic balance of synchronously improving conditions and factors	Creatively directed dynamic balance of conditions and factors of development at their synchronous improvement
Sustainable regressive development (form 2)	Anti-progressive development of the system, based on dynamic balance of dysfunctional realized and synchronously collapsing conditions and factors	Destructively directed dynamic balance of conditions and factors of development in their simultaneous destruction
Unsustainable development (essence)	Development of the system based on dynamic imbalance of conditions and factors	Conditions and factors are, at the same time, in a state of dynamic imbalance and change
Unsustainable progressive development (form 1)	Turbulent development of the system based on dynamic imbalance of conditions and factors that are in a state of continuous progressive but asynchronous improvement and interaction	Conditions and factors of development are in dynamic creative imbalance at their spontaneous interaction and improvement
Unsustainable regressive development (form 2)	*Turbulent development of the system based on dynamic imbalance of conditions and factors in a state of continuous (synchronous and/or asynchronous) destruction*	*Conditions and factors of development are in dynamic destructive imbalance at their spontaneous destruction*

are increasing [6, 9, 10]. Knowing "its place" in the general hierarchy of forms of sustainable development, it is possible to outline the immediate tasks of the country's development.

Thirdly, the most effective is a progressive form of sustainable development, involving a dynamic balance of constantly improving conditions and factors. This form, as a "system of project", is the desired goal of the respective efforts of the subjects of household, society, economy, environment and other elements of overall development. To achieve this goal, it is necessary to update the "system of object", consisting today of a triad: *society, ecology, economy.* Taking into account the requirements of endogeneity and inclusiveness in solving modern socio-economic problems, the object structure of the analysis of the problem of sustainable progressive development can include *the personality* and *the state.* Personality is necessary as an integral function of sustainable progressive development through the consistent development of a *worthy life.* A state that expresses the interests of civil society is necessary as a generator of *relevant institutional capital.*

References

1. Bobylev SN, Zubarevich NV, Solovieva SV (2015) Challenges of the crisis: how to measure the sustainability of development? Issues Econ 1:147–160
2. Buzgalin AV (2018) Sunset of neoliberalism (to the 200th anniversary of the birth of Karl Marx). Issues Econ 2:122–142
3. Buzgalin AV, Kolganov AI (2006) Man, market and capital in the economy of the XXI century. Issues Econ 3:125–142
4. Gegel GVF (1990) Philosophy of law. Thought, 524 pp
5. Drobyshevsky SM, Trunin PV, Bozhechkova AV (2018) Long-Term stagnation in the modern world. Issues Econ 11:125–142
6. Ershov MV (2017) Russia and the world: how sustainable is economic growth? Issues Econ 12:63–81
7. Zamulin OA, Sonin KI (2019) Economic growth: Nobel prize 2018 and lessons for Russia. Issues Econ 1:11–37
8. Kapelyushnikov RI (2015) the Idea of "age-old stagnation": three versions. Issues Econ 5:104–134
9. Kleiner GB (2015) Stability of the Russian economy in the mirror of system economic theory, Part I. Issues Econ 12:107–124
10. Mau VA (2018) At the end of the global crisis: economic problems of 2017–2019. Issues Econ 3:5–30
11. Polterovich VM (2018) To the General theory of social and economic development. Part 2. Evolution of coordination mechanisms. Issues Econ 12:77–103
12. Dictionary of philosophical terms. INFRA-M, 731 pp
13. United Nations (2012) The future we want, our common vision. Outcom Document of the 8 Rio+20 Conference. United Nations, N.Y
14. World Bank (2006) Where is the wealth of nations? Measuring capital for the 21st Century. Washington, DC
15. World Bank (2013) World development indicators. Washington, DC

Decomposition of the Object Structure and Goals of Sustainable Progressive Development

Irina S. Salikhova

Abstract The aim of the article is to concretize the object structure of sustainable progressive development. The relevance of the topic is due to the fragmentation of existing approaches to the characteristics of the structure of sustainable progressive development, in which there is no implementation of the requirements of the inclusiveness concept. Methodical basis is a structural-genetic approach to the problem, as well as the unity of qualitative and quantitative analysis. The main scientific results are the qualitative update and formation of the "five-link" object structure and relevant goals of the sustainable progressive development concept instead of the existing "triad".

Keywords Sustainable progressive development · Object structure · Civil society · Legal state · Institutional capital · Worthy life

1 Introduction

The existing paradigm of sustainable development in the world is effective enough, but not so much as to prevent the development of destructive processes that destroy not only the environment, but also the person himself. Despite the vigorous activity of the world community, represented by the UN, as well as national States in a way of forming a "green economy" and ensuring the equitable distribution of resources, the overall development trend is characterized by the predominance of negative trends over positive results. The complex factors of unsustainable development are poverty, lack of food and drinking water, environmental disasters, geopolitical ambitions of large countries and military conflicts, and a slowdown in production growth [1, 5, 17]. As a result, it becomes clear that the current ideology of sustainable development requires qualitative renewal.

I. S. Salikhova (✉)
Moscow Witte University, 12, 2nd Kozhukhovsky pas., Bldg. 1, Moscow 115432, Russia
e-mail: irinasalikhova@yandex.ru

A. V. Semenov and I. A. Sokolov (eds.), *Sustainable Development: Society, Ecology, Economy*, Earth and Environmental Sciences Library,
https://doi.org/10.1007/978-3-030-73110-6_5

A functioning paradigm has at least two drawbacks. First, its recommendations are not targeted, but universal in nature, within the framework of which the liberal-monetarist form of the General concept is realized, which is not the best among all forms. Secondly, this paradigm is characterized by a marked incompleteness of the actual structural approach to the problem, which is manifested in two aspects. The first aspect is the incompleteness of the object structure of the sustainable development analysis, which includes only society, ecology and economy [17, 18]. The second aspect is associated with the lack of reproductive vertical, designed to provide a systematic solution to the problem not only "left to right", but also "bottom-up" [6, 11, 15]. In this regard, the scientific hypothesis is the assumption that structural innovations in the study of sustainable development problems, first of all in the form of an in-depth decomposition of the "object system" [7], will provide a relevant shift of the current paradigm and soften the contradiction between theory and practice.

2 Methodological Matrix of the Study

The methodological and methodical capital of the current paradigm of sustainable development analysis is aimed, as noted, at the well-known "triad" in the form of society, ecology and economy. Local research programs are formed around the goals and objectives that ensure the sustainability of each of these elements, which are subsequently integrated into a single scientific and practical complex [6, 15, 17]. However, some important methodological and methodical aspects of the problem remain "behind the scenes". First, the existing set of goals and objectives is being implemented in the absence of a clear interdisciplinary understanding of the multiplicity of real conditions and factors of sustainable development. It is obvious that even within the framework of the mentioned "triad" the most effective methodological approach should be based on the post-non-classical analysis of the problem. This approach aims at a more vigorous use of relevant recommendations and conclusions of all sciences of man, society and nature exactly in the context of sustainable progressive development [8, 12, 16].

Secondly, the current paradigm does not take into account current trends in social development. First of all, such trends should include the imperative of inclusiveness in solving the problems of sustainable development, which aims, at the same time, at a universal and concrete solution to the problem [4, 9, 14]. In this regard, the next important trend is the growing need of man and society in a *new model of social state,* designed to transform economic turbulence into socio-economic sustainability on the basis of expanded reconstruction of relevant institutional capital [2, 9, 11]. Also, a special trend of modern development is the need for the most complete consideration of the civilizational features of the region. Attempts of implantation of civilizational values in the "alien soil" cause system instability.

Third, in order to ensure the specifics of solving the problems of sustainable development, the general declarative idea of the study should be transformed into a systematic and decomposition analysis of *sustainable progressive development.* Only in this case it is possible to count on the clarity of setting specific creative goals and the relevance of the mechanisms of achieving them. If sustainable development, in general terms, can be understood as the development, based on the dynamic equilibrium conditions and factors, the sustainable progressive development, it is reasonable to interpret as the development, due to dynamic equilibrium conditions and factors that are in a state of mutual simultaneous improvement. This interpretation actualizes the need for a new concept, focused on the comprehensive improvement of process and target forms [10, 11, 17].

Fourth, the analysis of dysfunctions of the existing paradigm of sustainable development, as well as the imperative of its progressive form, focuses on the need for further development, both horizontal analytical perspective and the formation of a new, vertical research algorithm. In this regard, and also taking into account the requirements of the principle of inclusiveness, the target function can be the sustainability of progressive human development on the basis of ensuring the sustainability of his *worthy life*. It is important to note, however, that the sustainability of a worthy life does not necessarily imply the sustainability of all other elements of the "object system". The task is to ensure that all these objects, often developing unevenly, but always progressively, as a result provide "final" sustainability in the form of progressive human development [2, 9]. Thus, the methodological matrix of the study determines the correct application of the unity of qualitative and quantitative analysis.

3 Discussion and the Study Results

Following the logic of the study, the key object in the framework of sustainable progressive development is the person, and the main function of this development is the sustainability of its progressive development through the steady provision of a worthy life [1, 12, 16]. The novelty of this analysis lies in the integration of the concept of anthropocentrism, general anthropology and the phenomenon of a worthy life into a new paradigm of sustainable progressive development. At the same time, the most acute aspect of the discussion is the provision that sustainable progressive development of a person as a function of ensuring his decent life may be the result of "unsustainable progressiveness" of other elements of the object structure, that are being taken into consideration. Moreover, "excessive perseverance" in achieving sustainable progress of all components of this structure can give the opposite result in the form of slowing down the progressive development of man, society and civilization. In particular, the metaphysical goal of "catching up with the world GDP growth rate" in the conditions of raw material economy can harm domestic natural capital, destroy the environment, "freeze" investments in fundamental science, weaken small business, catalyze social contradictions, etc.

Also, the novelty of the proposed study and the main systemic result is that in used object structure completed "vertical cycle" is clearly fixed, including the following elements. First, it is a person who has a certain level of implicit and explicit knowledge and who must independently ensure the stability and progress of his life. Secondly, the same person, striving for progressive development, creates a request for the sustainability of society, which is designed to recreate relevant cultural values and ideas. Third, such a society must be able to form a legal state (lawful state), which is called upon to broadly reproduce the relevant institutional capital. Fourth, this capital steadily forms progressive common vector of economic dynamics, not excepting certain cyclical fluctuations. Finally, the concept of progressive economic creation naturally determines the need to restore the environment as a general condition for sustainable human development.

It is obvious that the mentioned five-link object structure (person, society, state, economy, nature) of sustainable development requires some consistent specification. *The first,* it is unlikely that a person, striving for self-improvement, will remain within the neoclassical paradigm of "egoistic individualism". Now the sustainable progressive development of man should be linked with the system of his meanings (cultural values) and energetic ascent to his own creative and intellectual ecology (to creative personality destination). Socio-economic basis of sustainable progressive development of man, as well as the sustainability of his creative motivations, is a protected system of ownership of the person, functioning as an individual, as a member of various organizations and a member of society [13, 14].

The second circumstance is connected with the characteristic of sustainable progressive development of society. Socio-economic basis of sustainable progressive development of society is the system of common property. The function of sustainable development of the civil society quality integrity is its ability to recreate the legal state [2, 9, 11]. The imperative of the legal state appears as *the third* element of the object structure analysis of the sustainable progressive development concept. The need for this element is determined by the growing demand for relevant institutional capital. The legal system of ownership plays here an important role. The meaning of this was clearly expressed by G. Hegel: "justice is something great in civil society: good laws lead to the prosperity of the state, and free property is the main condition for its brilliance" [3, P. 264].

The fourth element of the object structure of the general concept of sustainable progressive development is the economy. By analogy, sustainable progressive economic development is economic growth based on a dynamic balance of relevant conditions and factors of production. In this regard, it is necessary to humanize the entire system of socio-economic indicators. In particular, the average GDP ("per head") should give the way to inclusive GDP ("reach everyone"), etc. In addition, the list of different variables and indicators should be interdisciplinary and "revolve" around the creative form of human capital. This includes the index of economic freedom, the level of protection of property rights and many other variables. An indicative list of such interdisciplinary indicators is presented in Table 1.

Fifth, the most important element of the object structure of sustainable progressive development is natural capital. Sustainable progressive development of nature is a function of the dynamic balance of its "living" and "inanimate" component with their simultaneous relevant improvement. In the future, such development should be accompanied by a complete cessation of the use of non-renewable natural resources. In relation to the economy, we are talking about the noospheric type of reproduction

Table 1 An approximate list of interdisciplinary indicators characterizing the stability of the domestic economy [19]

Indicators	2008	2010	2012	2014	2016	2017
The political rights index (1-max; 7-min)	6.0	6.0	6.0	6.0	6.0	7.0
Civil rights and freedoms index (1-max; 7-min)	5.0	5.0	5.0	6.0	6.0	6.0
The economic freedom index (100-max)	49.8	50.3	50.5	51.9	50.6	57.1
The property rights protection index (10-max; 0-min)	4.0	4.3	4.5	4.8	4.6	4.0
The happiness index (10-max)			5.46	5.72	5.96	5.81

[12, 16]. In this regard, the indicator of "economic capacity of the biosphere" is unlikely to contain useful research potential [15, 18]. Now the general innovative model of the object structure and the corresponding goals of sustainable progressive development is presented in Table 2.

4 Conclusion

Thus, a systematic confirmation of the proposed scientific assumption consists in the following conclusions. First, the challenges of sustainable and progressive development should not be simplified, but based on the extreme complexity of the problem. The conscious complication of the scientific and practical paradigm of analysis is the right way of the expected high efficiency of solving modern problems [8]. This means that the "object system" must have a large decomposition, both vertically and horizontally. In particular, the sustainable progressive development of the economy can be divided into the same development of households, firms, clusters, industries, regions, macro-regions, etc. The general conclusion is next: the more controlled and consciously formed elements of sustainable development of the system, the more chances to ensure its real progress.

Secondly, sustainable progressive development of the proposed innovative object structure *is not an end in itself.* Such development is designed to achieve sustainable and progressive human development as a function of a worthy life. The research value here is that a person acts as a "personified integral" of the whole system of social relations [4]. Man also appears as a product of evolution and a "clot of reason" of nature itself, which indicates that *human creativity cannot and should not be higher than the creativity of nature.* Here immeasurably increases the role of the legal state, designed to transform the "progressive turbulence" of the object structure relevant elements in a steady progressive development of the final result in the form of a worthy life and comprehensive human development. This is the key function of a new model of social state, which requires the relevant quality of the political and managerial elite, as well as highly efficient bureaucracy [2, 9]. At the same time, the state becomes an integral tool for the realization of the general need

Table 2 Innovative object structure and corresponding sustainable progressive development goals

The new "object system"	Qualitative integrity of the object of sustainable progressive development	Function of the object
Worthy human life	Formation of creative ideology of human development; recreation of humanistic values; activities in accordance with the imperatives of individual creative and intellectual ecology; uninterrupted self-development, etc.	Sustainable progressive human development and the demand for civil society
Worthy (civil) development of society	Improvement of the civic foundations of modern society: establishment of the "cult" of trust, justice, tolerance, creative activity and inclusive institutions; development of the creative elite, highly efficient bureaucracy and productive classes, etc.	Sustainable progressive development of society and the request for legal state
Worthy (legal) development of the state and creation of fair "rules of the game"	Development of the moral vector of the state ideology, which consists in the humanization of all forms of social relations; highly effective institutional capital; inclusive protection of property rights; development of highly effective bureaucracy; active reverse positive causal relationship of the legal state with civil society, etc.	Sustainable development of the legal state and request for high economic efficiency
Worthy being an economy, based on the noospheric type of production	A creative revolution that provides the economy with economically reproducible resources; ensuring the leading role of the economy of implicit knowledge and humanistically oriented industries; massive investments in the "growth points" of the latest noosphere technologies, consistently limiting the use of non-reproducible natural resources and factors, etc.	Sustainable progressive economic development and growing demand for noospheric natural capital

(continued)

Table 2 (continued)

The new "object system"	Qualitative integrity of the object of sustainable progressive development	Function of the object
Worthy being of the nature, based on the termination of natural potential destruction	Inclusion of the natural factor in methods of calculation of dynamic variables of social and economic development; priority development of investments for restoration of the destroyed natural potential; institutionalization of natural regeneration at the regional, macro-regional, national and supranational level, etc.	Sustainable progressive development of nature and growing demand for "human environmental"

for a reasonable regularity of man, society, state, economy and nature progressive development [12, 15].

References

1. Buzgalin AV (2018) Sunset of neoliberalism (to the 200th anniversary of the birth of Karl Marx). Issues Econ 2:122–142
2. Buzgalin AV, Kolganov AI (2016) Planning: potential and role in the market economy of the XXI century. Issues Econ 1:63–81
3. Gegel GVF (1990) Philosophy of law. Thought, 524 pp
4. Danilov YA (2019) Current state of the global scientific discussion in the field of financial development. Issues Econ 3:29–47
5. Drobyshevsky SM, Trunin PV, Bozhechkova AV (2018) Long-term stagnation in the modern world. Issues Econ 11:125–142
6. Ershov MV (2019) Ten years after the global crisis: risks and prospects. Issues Econ 1:37–54
7. Kleiner GB (2011) System resource of economy. Issues Econ 1:89–101
8. Lyubimov IL, Ospanova AG (2019) How to make the economy more difficult? The search for the causes of the complications. Issues Econ 2:36–53
9. Mau VA (2019) National goals and model of economic growth: new in the socio-economic policy of Russia in 2018–2019. Issues Econ 3:5–29
10. Mironov VV (2019) On the diagnosis of the current state of the Russian economy and medium-term prospects of its growth. Issues Econ 2:5–35
11. Polterovich VM (2018) To the General theory of social and economic development. Part 2. Evolution of coordination mechanisms. Issues Econ 12:77–103
12. Salikhov BV (2017) Quality management of creative capital in the noosphere knowledge economy: Monograph, ed "MU them. S.Y. Witte", 250 pp
13. Salikhova IS., Salikhov BV (2010) Decent human life as the basis of innovative development of modern economy. Innov Invest 3:37–49
14. Slobodenyuk ED, Anikin VA (2018) Where is the poverty line in Russia?. Issues Econ 1:104–128
15. Sustainable development: society, ecology, economy: proceedings of the XV international scientific conference; in 4 parts. "MU named after S.Y. Witte", 2019. Part 3, 518 pp

16. Tsaplin VS (2009) Reasonable, but not thinking. Moscow: Kraft + , 352 pp
17. United Nations (2012) The future we want, our common vision. Outcom Document of the 8 Rio + 20 Conference. United Nations, N.Y
18. World Bank (2013) World Development Indicators. Washington, DC
19. https://knoema.ru/atlas

Prospects for the Use of Renewable Energy Sources in the Territory of the Russian Federation

Alexander B. Dolgushin, Alexander A. Tsukanov, and Andrey D. Kondratiev

Abstract The role and trends in the development of renewable energy sources in global energy consumption is reflected in the paper. A detailed analysis of the impact of RES on the amount of CO2 emissions into the atmosphere is carried out. Also, the prospects for the use of alternative energy sources within the Russian economic system are determined.

Keywords Renewable energy sources · Generation · Alternative energy · Fuel · Environmental pollution · Conventional energy

Since the first industrial revolution, the needs of the economy and society in providing energy, either thermal or electricity, have been met mainly through the use of hydrocarbon fuels. However, over the past decade, renewable energy sources (RES) have been gaining an increasing share in world production. Thus, for 2019, the share of renewable energy sources (including hydropower) in the global energy balance increased on 1.1%, which amounted to about 27% of its total energy production (Fig. 1).

At the same time, it is important to note that the share of hydropower has remained virtually unchanged over the past decade. As a result, the main growth is accounted for by wind and solar power generation. Thus, according to the International Renewable Energy Agency (IRENA), global renewable capacities in 2019 increased at 7.4%,

A. B. Dolgushin (✉) · A. A. Tsukanov · A. D. Kondratiev
Moscow Witte University, 12, 2nd Kozhukhovsky Pas., Bldg. 1, Moscow 115432, Russia
e-mail: adolgushin@muiv.ru

A. A. Tsukanov
e-mail: atsukanov@muiv.ru

A. D. Kondratiev
e-mail: kdg040598@gmail.com

A. V. Semenov and I. A. Sokolov (eds.), *Sustainable Development: Society, Ecology, Economy*, Earth and Environmental Sciences Library,
https://doi.org/10.1007/978-3-030-73110-6_6

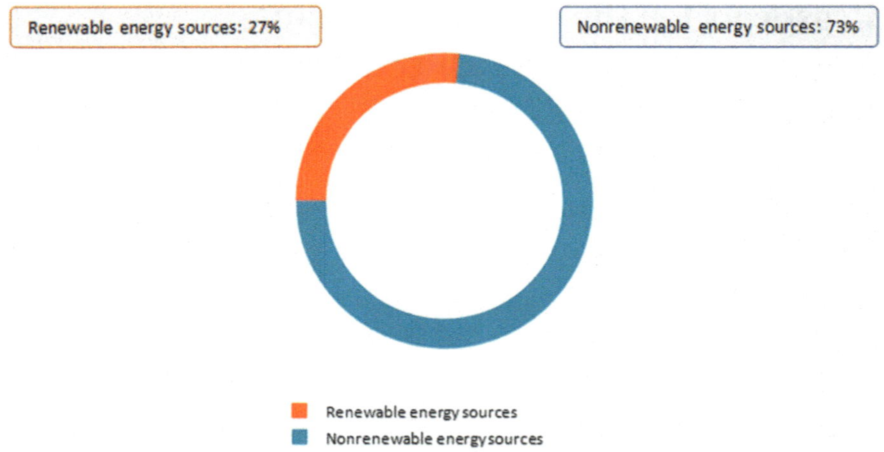

Fig. 1 The share of renewable energy sources in electrical power generation in 2019 (The official website of the World Energy Statistics Yearbook of the independent information and consulting company Enerdata – URL: https://yearbook.enerdata.ru (Date of the request: 25 July 2020).)

which is about 2,537 GW. At the same time, about half of this value is accounted for by solar and wind power generation.[1] The continuous decrease in the cost of solar and wind technologies has contributed to a significant increase in renewable capacity and electricity generation. Thus, in Germany, the share of electricity generated by solar and wind power plants amounted to 28.9% of the total volume (Fig. 2). Other European countries, such as Great Britain, Spain, Italy and Portugal, are not lagging behind in terms of indicators, where the average indicator of solar and wind energy generation in the total energy balance of these countries is 23.9%.

Such excitement among economically developed countries is caused primarily by the fact that during the period of active industrialization of society, which has been developing under the slogan of anthropocentrism, colossal damage was caused to the ecosystem, which entailed not only a decrease in the biodiversity of wildlife and climatic changes, but also an increase in diseases among the urban population. As for now, scientists have come to the conclusion that it is the combustion of traditional hydrocarbon fuels (oil, coal, gas), accompanied by abundant emissions of harmful substances into the atmosphere, that leads to the concentration of greenhouse gases and makes a significant contribution to the global warming trend. Thus, the production of heat energy and electricity, according to various estimates of scientists, accounts for up to 85% of total anthropogenic emissions of gaseous substances, in particular, combustion products of coal, gas, oil and fuel oil.[2] The response of the

[1] The Official website of the International Renewable Energy Agency IRENA – URL: https://www.irena.org/ (Date of the request: 25 July 2020).

[2] The official website of the "Ekobezopasnost" Companies Group – URL: https://ecobez.ru/zagrjaznenie-atmosfery-v-rezultate-szhiganija-iskopaemogo-topliva-neft-prirodnyj-gaz-ugol/ (Date of the request: 25 July 2020).

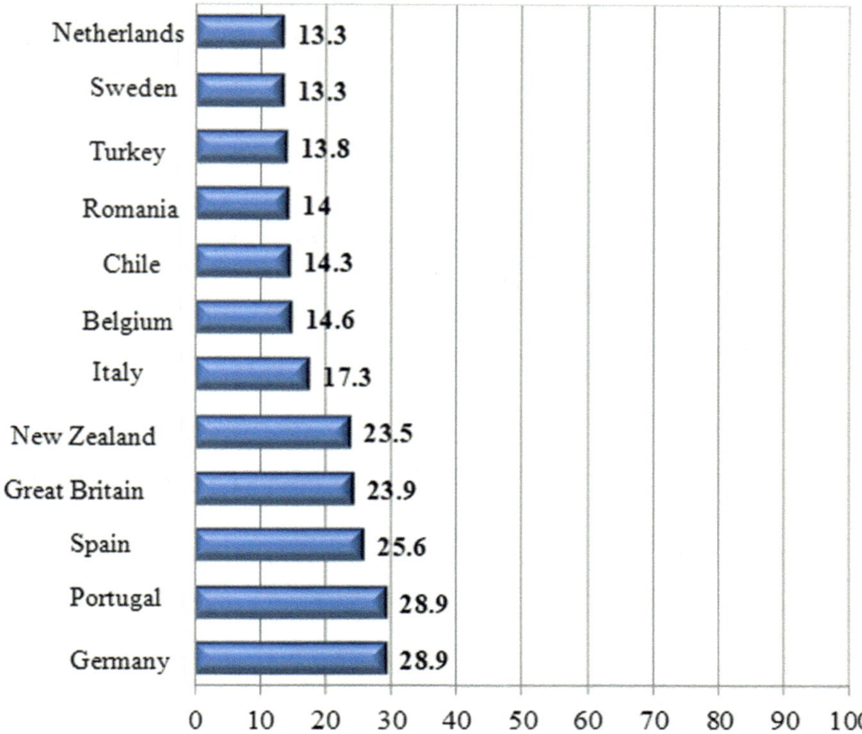

Fig. 2 Share of electricity generated by solar and wind power plants in relation to the total volume of electricity production in% (Compiled by the authors based on the World Energy Statistics Yearbook by the independent information and consulting company Enerdata 3. URL: https://ecobez.ru/zagrjaznenie-atmosfery-v-rezultate-szhiganija-iskopaemogo-topliva-neft-prirodnyj-gaz-ugol/ (Date of the request: 25 July 2020).)

world community to climate change was the creation of a number of international documents and agreements, such as the Kyoto Protocol, the Paris Agreement, etc. These measures have resulted in global CO emissions from coal being reduced on almost 200 mln. tons, or on 1.3%, compared to 2018 levels, offsetting the increase in emissions from the combustion of oil and natural gas. In advanced economies, emissions decreased by more than 370 mln. tons (or on 3.2%), with the electrical energy industry sector accounting for 85% of this reduction.[3] The largest decrease in emissions occurred in Europe, where the average annual growth rate of carbon dioxide emissions for the period from 2009 to 2019 was −0.958% (that is, about −37.1 Mt of average absolute increase in CO2 per year). In North America, renewable energy is also being actively introduced, especially solar and wind power plants, as

[3]The official website of the National Research University HSE – URL: https://energy.hse.ru/co2 (Date of the request: 25 July 2020).

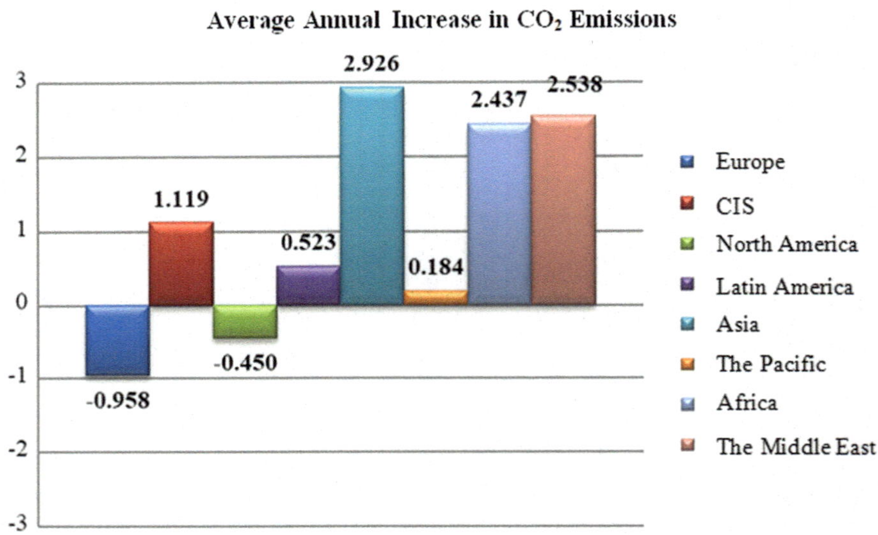

Fig. 3 Average annual growth rate of carbon dioxide emissions for the period from 2009 to 2019 in%. (Compiled by the authors based on the World Energy Statistics Yearbook by the independent information and consulting company Enerdata.)

evidenced by a decrease in the growth rate of CO2 emissions in this area over the past 10 years by −0.450% per annum (Fig. 3).

However, in Asia, CO2 emissions remain extremely high, about 15.879 Mt, which is on 33.426% higher than in 2009 and about 94% of the emissions from the rest of the world (or 48.5% of the global emissions). It is also worth noting the significant volume of emissions produced in Africa and the Middle East (Fig. 4).

Thus, according to the data presented in Fig. 4, it can be seen that the volume of CO2 emissions in these territories continued to grow throughout the entire analyzed period and amounted to 11% of the total volume of emissions produced by all other territorial entities, which represents 9.89% of the global emissions. Based on the data presented in Fig. 5, it can be concluded that despite significant progress in reducing CO2 emissions in Europe and North America, these achievements are completely covered by the growth of this indicator in Africa, the Middle East and Asia, the average value of which for this group of countries amounted to 29.6% relative to the values of 2009.

It is also important that in addition to a significant acceleration of ecosystem degradation, the consequences of fuel combustion have a direct impact on the health of the population. The main cause of such exposure is toxic microparticles from emissions that are inhaled by humans, which increase the likelihood of respiratory illness. For example, in China, where the share of coal generation is 67.5%, 670 thousand people die from coal combustion every year.[4] In India, where the coal is

[4] Annual report Congressional-executive commission on China. 2015. – URL: https://www.cecc.gov/publications/annual-reports/2015-annual-report (Date of the request: 25 July 2020).

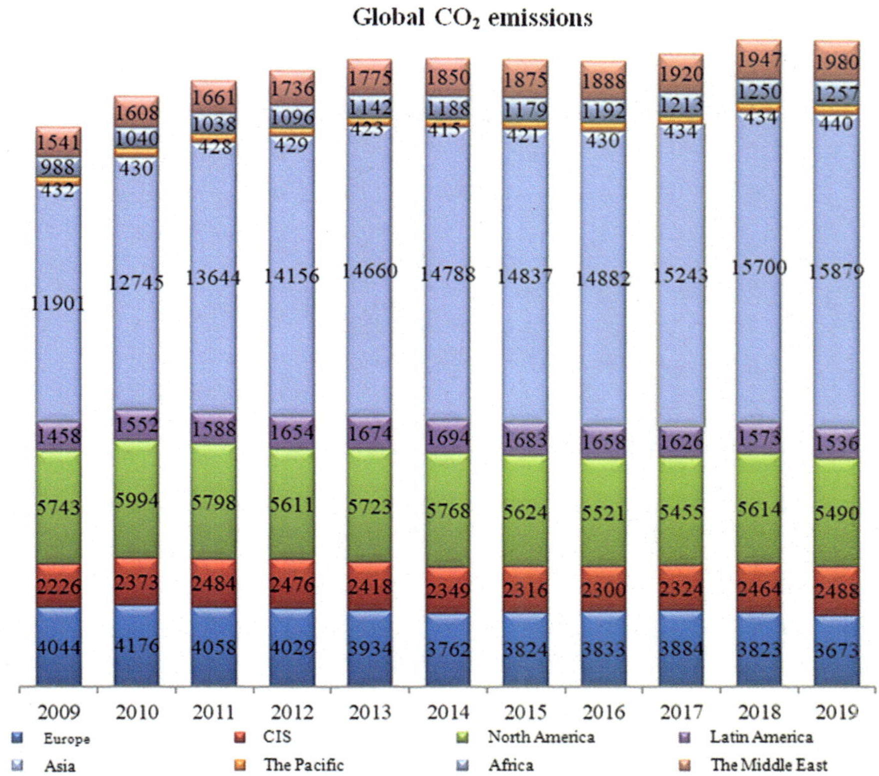

Fig. 4 Global CO2 emission for the period from 2009 to 2019, Mt. (Compiled by the authors based on the World Energy Statistics Yearbook by the independent information and consulting company Enerdata.)

also prevalent, this number is 115 thousand. It is also worth considering the growing needs of the population for energy associated with the rapid population growth, as a result of which the volume of consumption and production of fossil fuels, according to scientists' forecasts, will increase up to 50% by 2050.

As for Russia's integration into international green energy trends, it is worth noting that the capacity of the unified renewable energy system of our country, excluding large hydroelectric power plants, is less than 1%. Thus, the share of solar energy in the energy balance of the Russian Federation is 0.3% of the total installed capacity for 2019. However, in recent years, significant changes have been observed in this area. In 2018 alone, the increase in solar energy generators in the Russian Federation amounted to about 35% [1]. Such dynamism in the development of this area of green energy is due to the many advantages that stimulate the use of solar power plants in our country. First of all, this is the absolute inexhaustibility and abundance of the resource, environmental friendliness, as well as the fact that this type of generation is not accompanied by harmful emissions and has a wide range

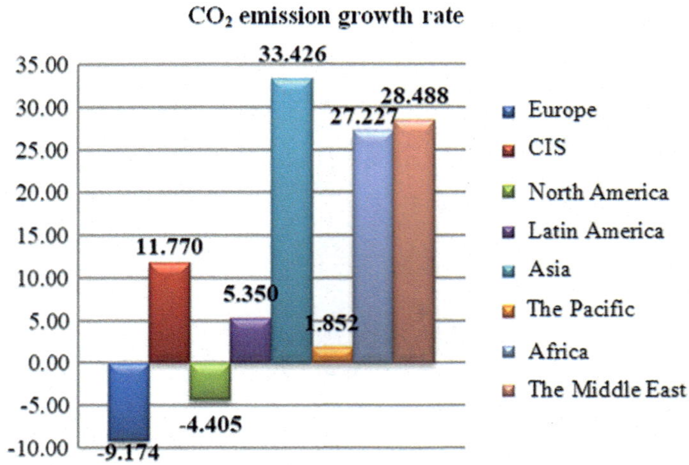

Fig. 5 The rate of CO2 emissions increase for the period of 2019 in relation to the value of 2009 in %. (Compiled by the authors based on the World Energy Statistics Yearbook by the independent information and consulting company Enerdata.)

of uses. It should be noted that the territory of Russia has enormous potential for the development of solar energy. The level radiated to the territory of Russia varies from 800 kWh/m^2 per year in the northern regions of the country to 1400 kWh/m^2 per year in the southern regions. However, with the active introduction of this technology, it is worth taking into account seasonal fluctuations that have a significant impact on the amount of solar radiation. Thus, at the same value of northern latitude, the level of solar radiation may differ by 6 times. For example, Isyangulovo SES, located in the Republic of Bashkortostan, in July 2018 generated 1,481 thsd. kW/h, and in January—only 198 thsd. kW/h of electricity.[5] However, despite certain shortcomings, it is safe to say that solar energy generation will take a leading position in the vanguard of the development of alternative energy in Russia. Thus, the volume of investments made only in solar energy from the total volume of all investments in RES has not dropped below 50% since 2017. In addition, solar energy has the greatest potential for cost reduction per unit of energy. Over the past 10 years, solar energy generation has fallen in price on 70%, which makes its further development even more attractive for the domestic economy.[6]

In general, at the present, the most promising from a technical, economic and environmental point of view is the supply of electricity to areas isolated from the unified power system. Thus, in the Republic of Sakha, which became the first constituent entity of the Russian Federation to adopt a law on renewable energy, power supply in most of the republic is decentralized. The same situation is typical for many regions

[5]The official website of the administration of the municipal region, Ishimbay region of the Republic of Bashkortostan – URL: http://www.ishimbaimr.ru/ (Date of the request: 25 July 2020).

[6]Global electricity review. March 2020. – URL: https://ember-climate.org/wp-content/uploads/2020/03/Ember-2020GlobalElectricityReview-Web.pdf (Date of the request: 25 July 2020).

of the Far East, such as the Magadan Oblast and the Chukotka Autonomous District. In total, at the moment, about 2/3 of the country's area remains non-electrified. According to various estimates, about 20 million people, who live in these territories, use autonomous generator of electric power, which runs by diesel, fuel oil and coal. According to Rosstat, up to 5 tons of oil products and up to 2 million tons of coal are supplied annually in the Far North. At the same time, due to unfavorable climatic conditions, the delivery of fuel to the place of consumption is rightfully considered laborious and economically ineffective. For the most distant consumers, the transport component in the total cost of the supplied fuel reaches 70–80% [2]. According to experts, if the potential for the development of renewable energy sources is fully realized in conjunction with measures to improve energy efficiency, then the cost of energy supply will be reduced on 40–50%, and the need for budget subsidies and the cost of paying bills for energy supply of budgetary organizations located in the regions of the Far North, will be reduced from 150–163 to 45–50 billion rub. per year. The share of budget expenditures in paying for energy supply in many regions of the Far North (without large industry) exceeds 30%, with an average level in Russia of 19.5% per annum. This share is the most significant in the Republic of Sakha (Yakutia) and is about 65%. Electricity tariffs for the population in isolated areas of the Sakha Republic vary between 3.83–5.47 rub./kWh. Tariffs for the population are about 5–25% of the economically justified level, which ranges from 17 to 205 rub./kWh [3]. That is why the "RusHydro" company is actively introducing renewable energy sources in this area. At the present moment, the company's Far Eastern arsenal has 19 solar power plants in the Republic of Sakha (Yakutia), as well as 4 wind power plants in the Republic, Kamchatka Krai and Sakhalin Oblast.[7] An important feature of these power plants is the developed automatic control system, which, if the insolation or the required wind speed is not enough for the requested energy generation level, switches the power plant to diesel generation. First of all, this is due to economic feasibility, since each kilowatt generated by solar panels saves diesel fuel, simultaneously reducing harmful emissions from fuel combustion. This is also evidenced by the programs developed by the Ministry of Energy of Russia for the development of power systems in isolated regions, which provide for the modernization of systems through the introduction of solar and wind hybrid power plants combined with a diesel engine.

However, it should be noted that the field of alternative energy in Russia is at the initial stage and is not viable in itself. That is why on January 8, 2019, the Government of the Russian Federation, basing on the priority nature of the development of renewable energy sources, decided to provide subsidies from the federal budget for state support for technological connection of generating facilities operating on the basis of the use of renewable energy sources.[8] Thus, in order to implement the policy

---

[7]PJSC "RusHydro" official website. – URL: http://www.rushydro.ru/ (Date of the request: 25 July 2020).

[8]The Ministry of Energy of the Russian Federation official website. – URL: https://minenergo.gov.ru/node/489 (Date of the request: 25 July 2020).

in the field of RES development, Federal Law No. 35-FZ of the Russian Federation, dated March 26, 2003 "On Electric Power Industry" provides for the use of a mechanism for selling the capacity of generating facilities operating on the basis of renewable energy sources (RES) under power supply contracts to the wholesale market (PSC RES).

Competitive tendering has also become widespread among the mechanisms for supporting RES. They represent a tender for the supply of a certain amount of electricity. Such projects are guaranteed a "green tariff", which allows investors to recoup their investments with a profit of 12% per annum for 15 years. In parallel with this, such projects are designed to stimulate the domestic production of components for power plants, which can become an important component for maintaining the trends in the promotion of Research and Development (R&D) in this area [4].

Of course, there is no question that alternative energy will be able to oust or at least significantly drive out traditional energy in Russia in the near future. Russia, as a country, has a solid foundation that makes traditional hydrocarbon energy feasible for a much longer period of time than in the most of the leading countries in terms of the share of RES in the energy balance. In Russian reality, the indicator of economic feasibility is a key element when considering the issues of introduction of alternative generation installations. Therefore, at the moment in Russia, to a greater extent, alternative energy plays the role of a reliable tool complementing traditional types of energy in supplying electricity to isolated regions of the Far East and the Far North, as in these regions there is no way to connect to centralized electricity generation and there are no good reasons to believe that this trend will change in the near future.

1 Conclusions

Despite the fact that many European countries are already actively introducing RES, Russia is only at the preparatory stage of this irreversible process. However, despite the lagging behind the world market leaders in the field of green energy, hybrid installations, which are actively developing in the Far East of the country, not only prove their economic efficiency and pay off investments, but also provide an additional effect that will significantly reduce government subsidies to support hydrocarbon (diesel, fuel oil, coal) energy in remote areas. However, this does not mean that the use of alternative energy sources in Russia will always be limited to an incomplete implementation. In reality, it is a matter of time before expansion into other energy supply niches. At the moment, supplying isolated areas of the country represents only a small share of the market. However, precisely this can awaken interest and make a significant contribution to the development of renewable energy in the country.

References

1. Sidorovich V. World Energy Revolution. How renewable energy sources will change our world/ V. Sidorovich - Moscow: Alpina Publ., 2016. - 208 p
2. Popov GE, Tulchinskaya YaI (2012) An Innovative approach to building of energy management system for remote settlements. "Oil and Gas Business" Electron Sci J 3:413–424. www.ogbus.ru. Accessed 25 July 2020
3. Bashmakov IA, Dzedzichek MG (2017) The assessment of energy supply costs in the regions of the far North. Energy Conserv 4:40–51
4. Barinova VA, Laitner DA, Lanshina TA (2020) Prospects for renewable energy development in Russia and the World. Federal State-Funded Educational Institution of Higher Professional Education "Russian Presidential Academy of National Economy and Public Administration", Moscow, 54 p. https://ssrn.com/abstract=2812577. Accessed 25 July 2020

Improving the Methodology for Selecting a Private Partner in the Implementation of Public-Private Partnership Projects

Aleksey S. Zhidkov

Abstract The problems of implementing public-private partnership projects that arise at the stage of its participants‘ bidding process are discussed in the paper. The world experience of pre-qualification selection is analysed, and the approaches of international organizations and states in determining the criteria for PPP contracting competitions are considered. Russian legislation is being analysed, which defines the method of pre-selection of participants, as well as the criteria to be applied to them, which are established by law. The negative experience with the concession agreement, which is based on an inefficient private partner selection process that does not involve the examination of a financial criterion, has led to the judicial termination of the concession contract and loss of budgetary resources, is being studied. It is argued that additional criteria are needed to improve the effectiveness of the pre-qualification of bidders for PPP contracting.

Keywords Public-private partnership · Concession · Bidding · Pre-selection · Evaluation criteria of the bidder

1 Introduction

The development of the public-private partnership institution in the Russian Federation has passed a long way of its formation. This form of interaction between government and business became especially active in the early 2000s, when the need to find additional sources of funding for the implementation of large infrastructure and socially significant investment projects was realized at the state level, since the limited budget funding did not allow these projects to be implemented on a socially necessary scale.

Until today, the approach to defining the essence of public-private partnership at the state level has been somewhat blurred. Thus, before the adoption of the Federal law "On public-private partnership, municipal-private partnership in the Russian

A. S. Zhidkov (✉)
Moscow Witte University, 12, 2nd Kozhukhovsky Pas., Bldg. 1, Moscow 115432, Russia
e-mail: ajidkov@muiv.ru

A. V. Semenov and I. A. Sokolov (eds.), *Sustainable Development: Society, Ecology, Economy*, Earth and Environmental Sciences Library,
https://doi.org/10.1007/978-3-030-73110-6_7

Federation and Amendments to Certain Legislative Acts of the Russian Federation" dated 13.07.2015 No. 224-FZ (hereinafter - Law No. 224-FZ), which came into force in 2016, each constituent entity of the Russian Federation had its own regional law which in one way or another regulated issues of public-private partnership and only with the adoption of Law No. 224-FZ at the Federal level, institutional rules and regulations in this area were defined. However, even today, the legislation in the field of PPP does not stand still and demonstrates active development, as evidenced by the regular amendments to the adopted law.

Thus, public-private partnership is becoming increasingly popular and interesting among regional and municipal authorities, especially with the acquirement of extensive experience in implementing investment projects based on this mechanism.

2 World Experience in Pre-qualification of Participants

The experience of PPP projects in the world community and attempt to critically reflect on the existing mechanism in the Russian reality with a view to finding possible ways of optimizing and improving it will be analysed in the paper.

Considering various approaches to defining "public-private partnership" concept, we can draw a certain conclusion that in international practice, PPP is understood as the interaction of the state and business to reach common goals, based on risk sharing and pooling resources.

In Russian legislation, there are two forms of such interaction - public-private partnership and municipal-private partnership (based on Law No. 224-FZ) and concession agreements (based on Law No. 115-FZ) [1, 2].

Prior to the adoption of Law No. 224-FZ, domestic specialists understood public-private partnership both as a concession, and as forms that go beyond the limits of concession agreements, however involve pooling resources and sharing risks between the state and the private party in the implementation of any joint projects. Legislative regulation in this case was carried out by regional laws on PPP. Almost every region had its own legislation or draft laws regulating the implementation of PPP projects, but with the adoption of Law No. 224-FZ, the "rules of the game" and the basic principles of interaction between PPP entities were defined. From now on, PPP in Russia officially refers to relations that are built and regulated by Law No. 224-FZ, as well as concession agreements based on Law No. 116-FZ [1] (Fig. 1).

The main attention in the paper is focused on the two mechanisms of interaction between authorities and business in improving the method of selecting a private partner in PPP projects - on the basis of the Law №224-FZ and on the basis of Law No. 115-FZ, since, as practice shows, the concession agreements are widespread in the implementation of projects in the service sector, while Law No. 224-FZ is used rather passively.

Let us analyze the current algorithm of PPP mechanisms implementation in order to develop recommendations for improving the methods of selecting a private partner in the implementation of projects based on it.

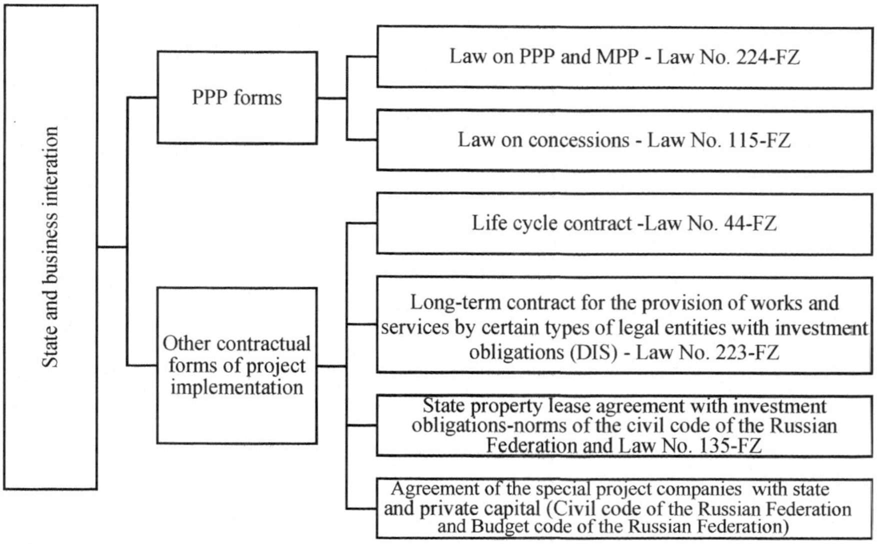

Fig. 1 Main forms of interaction between the state and business structures in Russia

Let's consider the main parties to the PPP and MPP agreements based on the current Russian legislation, presented in Fig. 2.

As noted earlier, the public partner and the private partner are the central parties to the PPP agreement.

A private partner is a Russian legal entity with which a PPP or a MPP agreement has been concluded.

Law No. 224-FZ establishes requirements for a private partner (part 8, article 5) [1].

Fig. 2 Main PPP subjects based on Law No. 224-FZ [1]

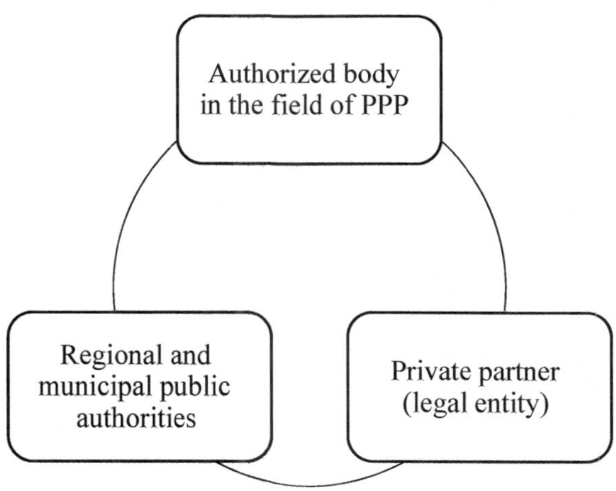

In addition to the private and state parties, the law defines authorized bodies in the field of public-private partnership (hereinafter – the authorized bodies).

The Federal law provides for mandatory identification of authorized bodies of the Russian Federation, constituent entities of the Russian Federation and municipalities.

Thus, for example, *the authorized body in the field of public-private partnership in Moscow* is the Department of economic policy and development of the city of Moscow [7].

The supreme executive authority of a constituent entity of the Russian Federation has the right to send a PPP project to the Federal authorized body (the Ministry of economic development of the Russian Federation) to assess the effectiveness and justify its comparative advantage.

Let us analyse the main stages of a PPP, MPP project consideration at the preliminary (project, documentary) stage, including the selection of a potential private partner for the subsequent conclusion of a PPP agreement.

According to the general rules, the consideration of a PPP, MPP project, includes 3 stages (see Fig. 3).

The algorithm for implementing the PPP and MPP project initiated by public authorities differs in that the development of the project is carried out by the state authorities themselves, after which they conduct a competitive procedure for the selection of a private partner.

In both cases, one of the fundamental stages in the implementation of PPP/MPP projects, as well as the concession agreement, is the *stage of selecting a private partner*. From a technical point of view, at this stage, the task of the state is to identify a suitable investor who would ensure the effective implementation of the project on the most favorable contractual terms for the public side.

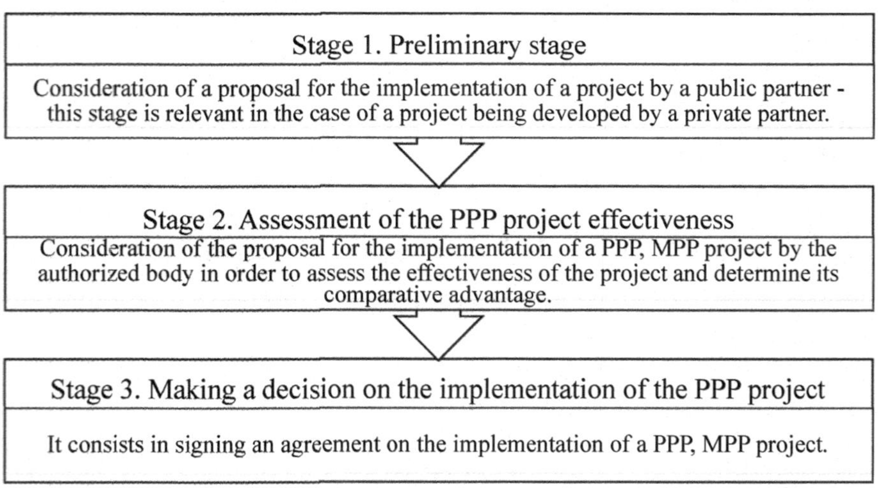

Fig. 3 Main stages of PPP (MPP) project implementation based on private business initiative [5]

Let us study the foreign experience of selecting a private partner to enter into a PPP/MPP agreement or a concession agreement.

As it has already been established, *the private partner selection stage* is one of the key elements of the PPP project implementation process. From a technical point of view, at this stage, the task of the state is to determine the appropriate investor who will ensure the effective implementation of the project on the most favorable contractual terms for the public side [5].

Analysis of the world practice of selecting a private partner in such countries as Brazil, Australia, Puerto Rico, Chile, Egypt, India, the European Union countries showed that there are two ways most common in this case:

1. Selecting a private partner based on direct negotiations with a single participant;
2. Conducting the bidding process among participants [9].

As a rule, the most effective way to select a private partner can be considered a bidding process, since it allows you to ensure competition among its participants, and choose the most effective and reliable partner, of course if the procedure is open, transparent, and the rules are strictly observed by all players, the participants themselves behave honestly towards each other.

Competition undoubtedly encourages the selection participants to offer the most profitable options, which helps to increase budget efficiency, *of course if the proposals put forward ultimately have specific opportunities for their implementation.* That is why open competitions are the most common mechanism for pre-selecting a private partner in the world practice.

The review of international practice shows that some countries, such as Puerto Rico and the European Union, have both direct negotiations and a competitive basis for selecting a private partner. In some countries, such as Brazil, the PPP law does not provide for direct negotiations at all [8].

Let's consider the key parameters of competitive procedures based on the world experience.

Competitive procedures have three main essential parameters that determine the format of selecting a private partner:

1. competition stageness;
2. formats for bids submission and evaluation;
3. possibility to adjust the terms of competition through negotiations with participants.

Let's consider the typical scheme of the competition, shown in Fig. 4.

Stages of the competition.

There are two main stages:

1. prequalification/pre-selection;
2. winner selection stage.

Thus, the essence of *pre-selection* is to weed out those contestants who do not meet a certain set of criteria and characteristics specified by the public side. The essence

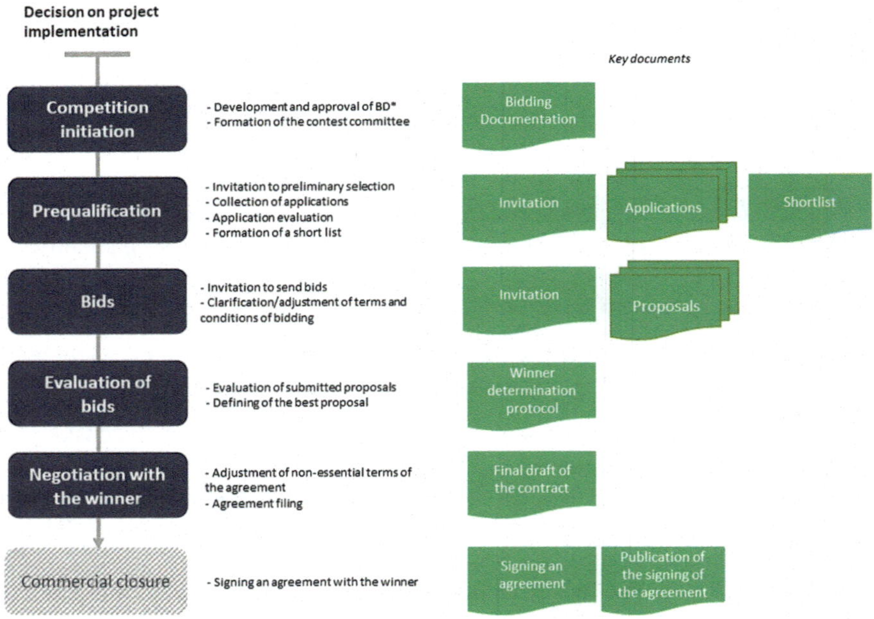

Fig. 4 Typical competitive procedure [11]

of the *winner selection* stage is a comparative analysis of the remaining participants in order to select the most favorable offer [10].

Thus, bidding may be:

1. single-stage (without prequalification);
2. two-stage (prequalification + winner selection).
3. two-stage (multi-stage winner selection without prequalification);
4. three or more stages.

Examples of the countries with different approaches to competitive events are shown in Fig. 5.

The data show that the most foreign countries pay great attention to pre-qualification selection of participants, to weed out those who are clearly unable to fulfill the terms of the agreement, which helps to repeatedly reduce the risks of non-implementation of the concluded PPP agreement.

Let us examine the features of prequalification based on the world experience in more detail. As you can see, this mechanism is used in most countries that practice PPP, as this stage allows the following [6]:

- selecting the most competent participants of the competition procedure, who will be able to provide the most realistic proposals in the future;

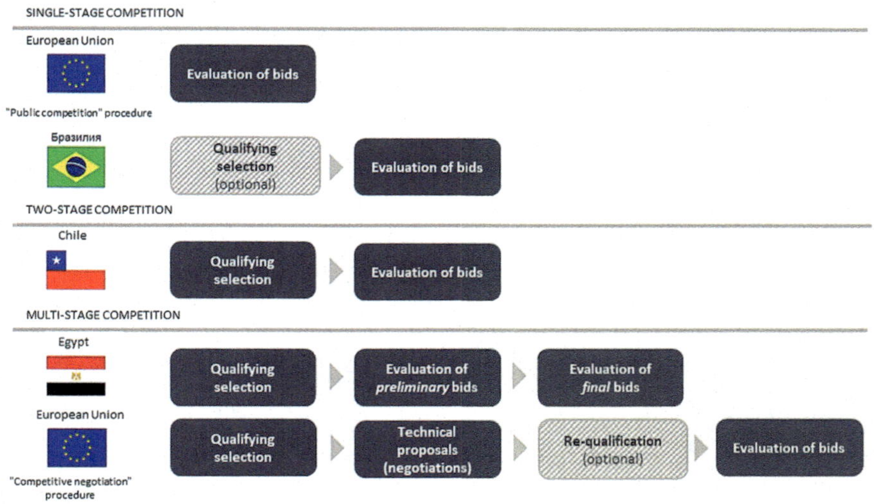

Fig. 5 Examples of the competitive procedures‘ stageness [11]

- reducing transaction costs associated with the subsequent analysis of a set of proposals, in case of admission of all willing private owners to the competitive procedure, without prequalification, and the time of decision-making on the conclusion of a PPP agreement is also decreased;
- due to the decrease in the number of participants in the competition and the increase in the chances of winning it, the motivation of the remaining participants to better and more thorough study of their proposal is growing.

However, there are certain negative aspects and risks when using prequalification. In particular, *a bad selection of criteria can lead to excessive restrictions on the number of admitted companies and reduction of further competition.* Also in case of insufficient experience in implementing PPP projects in general, which is specific for developing countries, this stage can lead to elimination of almost all potential participants.

In the world practice, it is customary to select an average of 3 to 6 participants within the prequalification. In some countries, for example in India, the number of participants is strictly limited to 6, in some countries the number of participants in the competitive procedure is not limited, for example in Singapore, and in some countries the minimum required number of participants is fixed, for example in the European Union there must be at least three.

Let’s consider the ***main criteria*** put forward by the participants of the competition in international practice.

The analysis of the pre-qualification selection experience in foreign countries showed that the effectiveness of this procedure is largely determined by what criteria were selected and what parameters of a potential participant were evaluated. Thus, we can distinguish two enlarged groups of criteria:

1. "formal criteria";
2. "meaningful" criteria [10].

The "formal" criteria are those that are independent on the conditions of a particular project. These criteria *are mainly aimed at assessing the status of the participant*. For example, whether the participant meets any formal legal requirements, the reliability of the participant is assessed on the basis of the presence or absence of legal claims, whether a bankruptcy or liquidation procedure has been launched against him, whether there is a history of default, etc.

The "meaningful" criteria are those that are aimed at assessing the participant's potential to effectively implement a specific project, thus they primarily depend on the conditions of its implementation. In international practice, such criteria are divided into *technical and financial* ones.

It should be noted that in different countries, approaches to the definition of criteria differ. Thus, in those countries where the number of participants is not limited, qualitative selection criteria are used to determine whether this participant meets them, for example, whether there is experience in implementing projects in this area.

In countries where the number of participants is limited to those who have successfully passed the selection, quantitative criteria with a certain weight can be used to rank the remaining participants. However, as a rule, both qualitative and quantitative criteria are used.

3 Specificity of Pre-selection of the Competitive Procedure Participants in Russia

The following are some specific examples of the pre-selection criteria used.

In the Russian Federation, the selection of a private partner within both PPP forms, the regulation of competitive procedures is generally identical, however there are differences in certain aspects.

Thus, the legislation of the Russian Federation provides for two-stage competitive procedures for the selection of a private partner in PPP projects, including the preliminary selection and evaluation of competitive applications. At the same time, the Law on PPP of the Russian Federation - in contrast to the Law on Concessions of the Russian Federation - establishes the possibility of excluding pre-selection from the procedure.

The selection of a private partner may also be carried out without competition, in particular in cases where only one successful participant entered the competition or, within the framework of the private initiative mechanism, no alternative proposals were received.

Let us consider the specificities of the competition in the Russian Federation in terms of *pre-selection of its participants.*

Pre-selection of participants is carried out by the competition commission in accordance with the procedure established by the competition documentation. The task of pre-selection is to determine the following:

- compliance of the application for participation in the competition with the requirements contained in the competition documentation.
- applicant's compliance with the requirements for the competition participants.
- applicant's compliance with the requirements for a private partner in accordance with the Federal law under which the PPP project is implemented.

On the basis of the results of the pre-selection of contest participants, the competition commission makes a decision on the admission of the applicant to participate in the competition or on the refusal on admission for the applicant to participate in the competition and draws up this decision by the Protocol of the pre-selection of the competition participants.

3.1 Pre-selection Criteria in the Russian Federation

In the framework of pre-selection, first of all, compliance with the "***formal***" criteria explicitly established by law is checked.

- the absence of a decision on liquidation or termination of the competition participant's activity;
- the absence of a decision declaring the applicant bankrupt and opening bankruptcy proceedings against a participant [3].

For agreements in the framework of the Law on PPP of the Russian Federation, additional basic requirements are also established, namely:

1. non-application of administrative punishment on the day of submission of the application for participation in the competition;
2. no arrears on taxes and fees not earlier than one month before the date of submission of the application for participation in the competition;
3. availability of licenses required in accordance with the legislation of the Russian Federation for certain types of activities [3].

In addition to the above-mentioned "***formal***" criteria, pre-selection may also include requirements for the qualifications, professional and business qualities of the participants. In fact, the list of such criteria in both laws is open.

Thus, summing up the results of the analysis of international and domestic practice for selecting a private investor, we can conclude that in foreign countries there is a serious emphasis on pre-selection of participants in the competition for concluding a PPP agreement. The main purpose of this selection is to determine the real possibility of implementation of the obligations assumed by the project being implemented by a private investor, which contributes to dropping out of the participants who submit to the competition attractive from a financial point of view documentation, but do not have the ability to achieve the declared results.

In this area, the Russian practice has serious drawbacks, since the preliminary selection of competition participants is carried out formally, based on the provided documentation without a detailed acquaintance with a specific applicant and an assessment of their potential material, technical and financial capabilities to fulfill their obligations, which often leads to the failure of the concluded agreement and is fraught with loss of state budget funds, which is confirmed by a lot of statistical data.

The drawback revealed as a result of the analysis requires the development of criteria that are adequate to the Russian reality and practice of implementing PPP projects, as well as the improvement of the procedure and methods for their selection.

Since the essence of the preliminary assessment of the competition participants today is reduced to their formal compliance with the existing bureaucratic criteria, in practice this leads to an unplanned negative result of the project.

Let us consider an example when the concession agreement was approved, but then terminated, due to non-performance of the concessionaire's duties.

Thus, in September 2011, a concession agreement was concluded between the "Ulyanovsk district" Municipality represented by the municipal institution " Committee on Managing Municipal Property and Land Relations of the "Ulyanovsk District" Municipality" and CJSC "BARS" in respect of facilities for the production, transmission and distribution of electric and thermal energy.

The concessionaire was also obliged to invest funds in the reconstruction of facilities in a total amount of at least 99.1 mln. rubles, inter alia to complete the reconstruction and hand over in operation several facilities by attracting investments in the amount of at least 30 mln. rubles by March 2012 [10].

In 2017, the Municipal entity "Ulyanovsk district" represented by the Municipal institution of the Administration of the municipal entity "Ulyanovsk district" of the Ulyanovsk oblast appealed to the Arbitration court of the Ulyanovsk oblast with a claim to CJSC "BARS" for termination of the concession agreement dated 02.09.2011 and the additional agreement dated 26.04.2013 to the concession agreement dated 02.09.2011 due to violations of the concession agreement, namely:

- violation of the established reconstruction (modernization) terms of the object of the agreement;
- use (operation) of the objects for the purposes not established by the agreement;
- violation of the procedure established by this agreement for the management of the object of the agreement, the use (exploitation) of the facilities;
- non-performance or improper performance by the concessionaire of their obligations.

The arbitration court of the Ulyanovsk region, having studied the materials, satisfied the claim.

Analyzing this situation, we can conclude that at the time of the conclusion of the concession agreement, CJSC "BARS" successfully passed the competition for the concessionaire selection and, therefore, fully satisfied the established criteria for the selection of a private investor.

Table 1 Financial statements of CJSC "BARS" for 2012–2017

YEAR	BALANCE, thsd. rubles.	REVENUE, thsd. rubles.	PROFIT, thsd. rubles.
2017	102 097	26 380	319
2016	94 486	59 872	494
2015	45 326	24 988	295
2014	33 484	14 674	170
2013	36 076	16 918	116
2012	22 464	19 390	306

Source https://synapsenet.ru/searchorganization/organization/1037301156463-ao-bars

Meanwhile, the considered financial indicators of CJSC "BARS" (OGRN-1037301156463), which was to invest at least 99.1 mln. rubles in the reconstruction of real estate, as well as to engage in the production, transmission and distribution of electric and thermal energy under the terms of the agreement, was making 306 thsd. rubles (see Table 1) of net profit and had 4 employees for 2012.

As a result, after the conclusion of the concession agreement, the CJSC "BARS" management simply could not implement it. However, this situation could have been avoided if at the stage of the participants' selection this organization had not been missed, taking into consideration its material, technical and financial and economic potential.

4 Conclusions

In conclusion, it should be noted that in order *to solve the identified problem, an event aimed at improving the current methodology of the competitive selection procedure at the prequalification stage is expected.*

Taking into account the considered international experience, it is necessary to enhance the preliminary assessment of the participants of the competition for the concluding a PPP agreement by introducing additional criteria that allow assessing the experience of the organization in the field of concluding a PPP contract, its material and technical base and financial and economic condition.

We would like to note that the existing criteria allow us to conduct the minimum necessary assessment of the participant's reliability and business reputation, however, they do not allow us to assess its experience, material and technical potential and financial and economic condition.

Therefore, it is evident that it is necessary to tighten the stage of selecting the competition participants through the use of additional selection criteria allowing to further assess the expertise, logistical base, financial and economic potential participant of the competition procedure.

The development of these criteria requires additional research since they have to consider both the features of the information collected about the participant's activities, and the characteristics of their economic activities in various sectors of the economy, where PPP mechanisms are widely used.

References

1. The Federal Law of 21.07.2005 N 115-FZ, "On concession agreements"
2. The Federal Law of 13.07.2015 N 224-FZ, "On public-private partnership, municipal-private partnership in the Russian Federation and Amendments to Certain Legislative Acts of the Russian Federation"
3. Resolution of the Government of the Russian Federation № 1322 of 04.12.2015 "On approval of the rules for pre-selection of participants in the competition for the right to conclude an agreement on public-private partnership, an agreement on municipal-private partnership"
4. Order of the Ministry of economic development dated 30.11.2015 No. 894 "On approval of the Methodology for assessing the effectiveness of public-private partnership project, municipal-private partnership and determination of their comparative advantages" (Registered in Ministry of justice of Russia on 30.12.2015 No. 40375)
5. Andreeva ES (2013) Assessing the effectiveness of public-private partnership projects: a methodological approach. Sci J "Bulletin of the Technological University" (Kazan National Research Technological University). 12. https://cyberleninka.ru/article/n/otsenka-effektivnosti-proektov-gosudarstvenno-chastnogo-partnerstva-metodologicheskiy-podhod. Accessed 11 Jan 2019
6. Krykanov DD (2019) Assessing feasibility of public-private partnership projects: analysis of Russian and International Practice. Public-private partnership, vol 2. https://cyberleninka.ru/article/n/otsenka-osuschestvimosti-proektov-gosudarstvenno-chastnogo-partnerstva-analiz-rossiyskoy-i-mezhdunarodnoy-praktiki. Accessed 03 Jan 2019
7. Makarov IN (2015) Methodology for assessing the effectiveness of public-private partnership projects in regional infrastructure. Public-private partnership, vol 1, no 1, pp 37–50
8. Assessing cost effectiveness in PPP projects. [Electronic resource]. http://www.oecd.org/gov/budgeting/45038620.pdf. Accessed 02 Nov 2018
9. Analytical Study Assessing Effectiveness of PPP Projects in Comparison with Other Forms of Infrastructure Projects Implementation in EAEU Member States. Research in the framework of the implementation of the Joint activities plan for 2017 within the framework of the Memorandum of cooperation on the development of public-private partnership in the member states of the Eurasian Economic Union. http://pppcenter.ru/assets/files/issledovanie_100118.pdf. Accessed 02 Nov 2018
10. Law on Public-Private Partnership: guide to application. Special edition for the international investment forum "Sochi-2015". http://pppcenter.ru/assets/docs/Zakon-Block_28-09-2015_v01.pdf. Accessed 02 Nov 2018
11. Private Initiative in Concessions: International Experience and Prospects of its Development in Russia, Moscow. http://pppcenter.ru/assets/docs/Initiative-A4-Block_20-04-2015_web++.pdf. Accessed 02 Nov 2018

Conceptual Framework for Managing the Competitiveness of Socio-Economic Systems

Aleksandr A. Ryazanov

Abstract The results of a study conducted by the author on the essence and content of the economic categories "economic competition", "competitiveness of the socio-economic system", "management of the competitiveness of the socio-economic system" are presented in the paper. Based on the systematization, generalization and critical understanding of the key provisions of the concepts of economic competition and competitive advantages, the levels and socio-economic functions of economic competition, the essential properties of the competitiveness of the socio-economic system are identified, the definitions of competitive advantage, competitive weakness, competitive potential, competitiveness of the socio-economic system are clarified, competitive opportunities and competitive threats of the external environment of the socio-economic system, as well as the systemic relationships between these categories, a model of the process of managing the competitiveness of the socio-economic system has been developed.

Keywords Economic competition · The competitiveness of the socio-economic system · Competitiveness management of socio-economic system

1 Introduction

The intensification of competition is the main challenge not only to sustainable development, but also to the existence of socio-economic systems of various levels in modern economic conditions. In this regard, one of the most important tasks of the controlling subsystems of socio-economic systems is the permanent increase in the competitiveness of the latter through the management of this complex characteristic.

It should be noted that the categories of "competitiveness" and "competitiveness management" have been the subject of research in economic science for many years, however a generally accepted approach to their essence and content by the scientific

A. A. Ryazanov (✉)
Moscow Witte University, 12, 2nd Kozhukhovsky Pas., Bldg. 1, Moscow 115432, Russia
e-mail: aryazanov@muiv.ru

A. V. Semenov and I. A. Sokolov (eds.), *Sustainable Development: Society, Ecology, Economy*, Earth and Environmental Sciences Library,
https://doi.org/10.1007/978-3-030-73110-6_8

community has not yet been developed, which makes it difficult to use the achievements of science in the practice of managing socio-economic systems. In this regard, the issue discussed in the paper seems to be relevant to the author.

The aim of the study is to solve the scientific problem of developing and substantiating conceptual provisions for managing the competitiveness of socio-economic systems at various levels.

Achieving this aim involves solving a set of interrelated tasks: to clarify and expand the conceptual and terminological apparatus of scientific research in the field of competitiveness management of socio-economic systems on the basis of systematization, generalization and critical reflection of key provisions of the concepts of economic competition and competitive advantages; to develop a model of the competitiveness management process of a competing socio-economic system.

The theoretical and methodological foundations of managing the competitiveness of socio-economic systems were laid down in the works of A. Smith, D. Ricardo, J. S. Mill, A. Marshall, W. S. Jevons, F. I. Edgeworth, P. Sraffa, E. G. Chamberlain, J. M. Keynes, J. V. Robinson, C. J. Arrow, J. A. Schumpeter, F. A. Hayek, G. A.Saimon, J. J. Stigler, I. M. Kirzner, R. G. Coase, M. E. Porter [1–18] and were developed in the works of many foreign and domestic experts. However, their researches do not cover all aspects of this multilevel and multifaceted phenomenon.

2 Materials and Methods

In the process of solving the main tasks of the study, such methods of studying social and economic phenomena as dialectic, systemic, historical-logical, comparative-analytical, expert assessments, as well as methods of scientific abstraction, induction and deduction, normative and positive analysis and synthesis were used.

3 Results and Discussion

The basis of the modern theory of economic competition is a set of its concepts, focusing the attention of researchers on certain aspects of this complex and multifaceted phenomenon. The analysis of the most widespread behavioral, functional, institutional, managerial and structural concepts in the scientific community allows us to draw a number of conclusions.

Economic competition is a structure-forming type of competitive behavior and competitive relationships of market participants. Competitive relationships of socio-economic systems make up a significant part of market relations, perform a number of important socio-economic functions (stimulating, differentiating, sanitizing, pricing, regulating, innovative, the function of supporting macroeconomic sustainability, allocative, adaptive, distributive, controlling, etc.) and are the key factor in the development of a market economy.

The nature and results of the competitive relationships of socio-economic systems largely determine the state of the respective markets and, therefore, serve as an evaluation criterion for identifying several models of market structures: the market of perfect competition, the market of monopolistic competition, oligopoly, the market of pure monopoly, etc.

Competitive behavior is typical of socio-economic systems of various levels (national, regional, sectoral, sub-sectoral economic complexes, corporations, economic clusters, enterprises and households), and also manifests itself at the commodity level.

Competitive behavior is the rational conflict behavior of participants in competitive relationships. It is based on a conflict of their contradictory economic interests. The system of motives for competitive behavior includes both motives for the most complete satisfaction of one's own needs, and the desire for victory over competitors, as well as for public recognition of this victory. In addition, sellers have a motivation to meet the needs of consumers of their products to the fullest extent possible. The quantitative certainty of the purpose of competitive behavior is given by the share of the corresponding competitive market occupied by the participant in the competitive relationship.

Considering the competitive behavior of the socio-economic system not as a behavioral one but as a social and economic category, its aim should be to redistribute in its favor the value created in the social and economic system of a higher-order in a given period and reaching better indices of economic and financial development than competing systems.

The competitive behavior of a socio-economic system of any level is a special case of the management process, the basis of which is the strategic component (strategic competitive goal-setting, strategic competitive analysis, selection, development and implementation of a competitive strategy, as well as strategic control). As for the functional and operational components of this process, they are subordinate in nature and solve the problems of marketing and resource support of the main stages of strategic management.

The subject of this management process is the management subsystem of the socio-economic system, as well as the management subsystems of socio-economic systems of a higher order, the element of which this system is. Its objects include competitiveness, competitive potential, competitive advantages and competitive disadvantages of this system.

It should be noted that most researchers recognize the determining impact of the competitive advantages of competing socio-economic systems on the course and outcome of their competition, as well as considering these advantages, along with competitive disadvantages, as the main element of the competitiveness of socio-economic systems. However, the scientific community has not developed a generally accepted approach to the essence of the competitive advantage of the socio-economic system yet. Proponents of several existing concepts of competitive advantage focus on their sources. The definitions they propose reflect only certain essential properties of this complex multifaceted category: specificity of manifestation, relativity, multi-leveledness, partial latency, certainty, integration, multi-parameterity, dynamism, controllability.

On the basis of a synthesis of these properties and the objective function of the competitive advantage of the socio-economic system, the author considers it possible to define it as a time-sensitive and manageable set of characteristics of the socio-economic system, superior the similar set of characteristics of competing social and economic systems and providing a potential opportunity for this system to redistribute the value created in the social and economic system of a higher-order economic system in a given period in its favor and also reaching better financial and economic development indicators than those of competing systems.

The modern theory of competition which only recognizes the fact of their existence pays unacceptably little attention to the competitive weaknesses of socio-economic systems.

The competitive disadvantage of the socio-economic system is the antipode of its competitive advantage and, therefore, can be defined as subject to temporal dynamics and a manageable set of characteristics of the socio-economic system, inferior to a similar set of characteristics of at least one of the competing socio-economic systems and providing a potential opportunity for redistribution of the value created in the socio-economic system of a higher order for a certain period in its favor, as well as reaching better than this system financial and economic development indicators for at least one of these systems.

Competitive advantages and competitive disadvantages of the socio-economic system can take various forms and belong to different types. At the same time, the author considers it not entirely correct that some researchers divide competitive advantages by the source of their formation into internal and external ones. The author believes that competitive advantages (as well as competitive disadvantages) of a socio-economic system can be formed only by elements of its internal environment. The external competitive advantages highlighted by some researchers, formed by elements of the external environment of the socio-economic system, in the author's opinion, should be attributed to the competitive capabilities of the external environment of the socio-economic system - environmental factors that contribute to the redistribution of the value created in the socio-economic system of a higher order for a certain period in its favor, as well as reaching the better financial and economic indicators of development than those for competing socio-economic systems.

Obviously, there are environmental factors that have an opposite effect on the system. In this regard, under the competitive threats of the external environment of the socio-economic system, the author proposes to understand the factors of the external environment that contribute to the redistribution of the value created in the socio-economic system of a higher order for a certain period in its favor, and also reaching the better financial and economic development indicators by one of competing systems which compete with this system.

Each competing socio-economic system at each moment of time has a certain unique set of competitive advantages and competitive disadvantages As noted above, the competitive advantages and competitive disadvantages of the socio-economic system act as the objects of the managing action of its managing subsystem, as well as the managing subsystem of the socio-economic system of a higher level. The target setting of this managing action is the identification, accumulation, preservation, development and implementation of competitive advantages, as well as the identification and elimination (compensation) of competitive disadvantages. In addition, they affect the elements of the internal environment of the socio-economic system, stimulating the formation, acquisition and transfer of the competitive advantages of the system.

The author considers it possible to determine a set of competitive advantages, competitive disadvantages, as well as elements of the internal environment of the socio-economic system, which form its competitive advantages and competitive disadvantages as the competitive potential of the socio-economic system, that is unique at each moment in time. This set can be characterized by both the predominance of competitive advantages over competitive disadvantages (positive competitive potential) and the predominance of competitive disadvantages over competitive advantages (negative potential), as well as their approximately the same ratio (zero potential).

Along with the elements of the competitive potential of a competing socio-economic system, its competitive behavior determines the prerequisites for realizing this potential. They are based on the aforementioned competitive opportunities and threats from the external environment. The combination of elements and prerequisites for realizing the competitive potential of the socio-economic system forms the basis of the latter's competitiveness. The author considers it possible to define the competitiveness of a competing socio-economic system as a dynamic, controllable property of a given system, determined by competitive potential that is not inferior to competing systems. As well as the possibility of its implementation in the process of competitive interaction at a given time, expressed in the system's ability to redistribute of the value created in the socio-economic system of a higher order for a certain period in its favor recognized by the environment. As well as reaching the better financial and economic indicators of development than those for competing socio-economic systems.

The conceptual model of the process of managing the competitiveness of the socio-economic system developed by the author is shown in Fig. 1.

This model unites subjects, targets, directions, objects, as well as stages of the process of managing the competitiveness of the socio-economic system and, on the author's opinion, fully reflects the essence and content of the latter.

4 Conclusions

The author believes that the provisions, generalizations and conclusions formulated in this paper allow a deeper study of the essence, content and interrelationships of economic competition, competitiveness, competitive potential, competitive advantages, competitive disadvantages of competing socio-economic systems, competitive

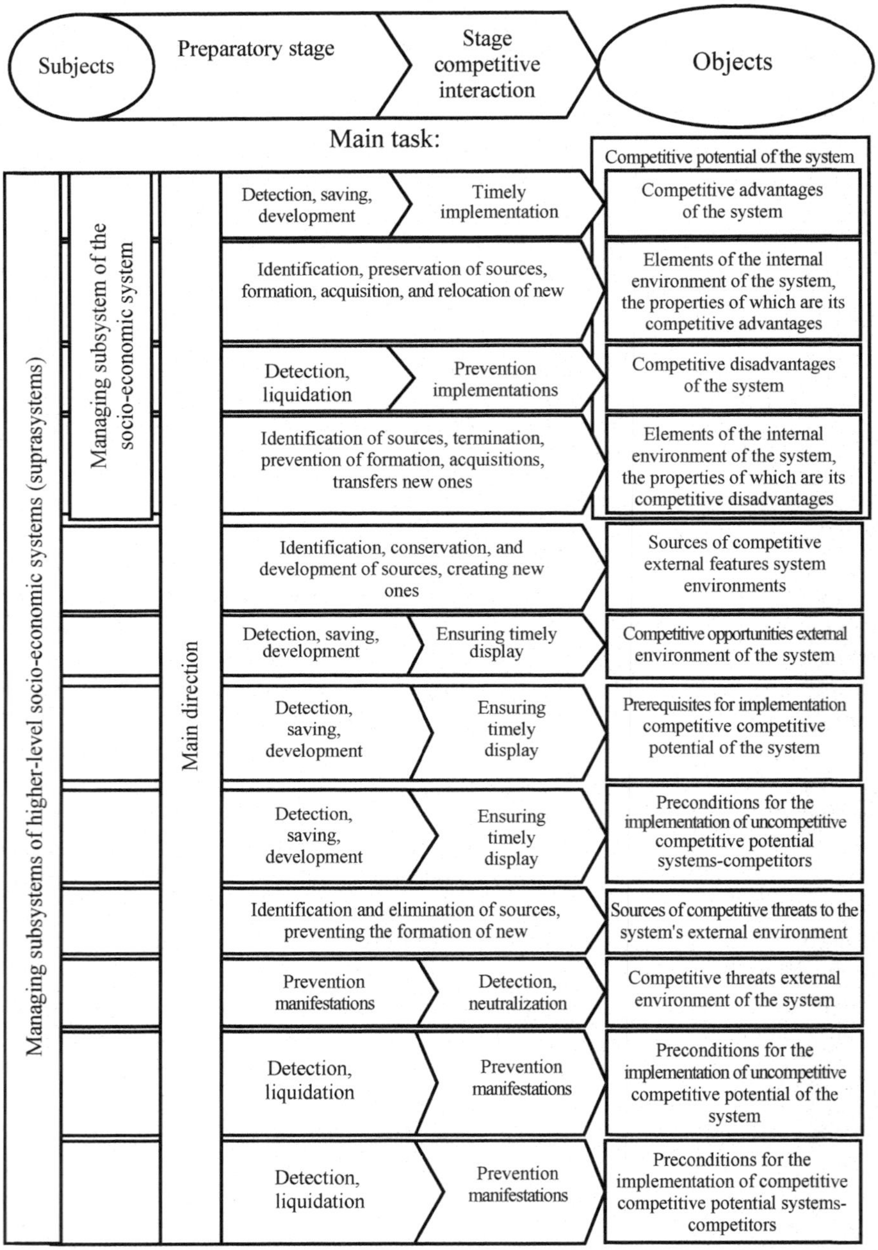

Fig. 1 Conceptual model of the process of managing the competitiveness of the socio-economic system

opportunities and competitive threats of their external environment, as well as the process of managing the competitiveness of the socio-economic system and can serve as the basis for the formation of a modern concept of the latter.

In addition, the results of this study can be taken into account and used in the practical activities of officials and governing bodies of socio-economic systems at various levels.

References

1. Smith A (2016) An inquiry into the nature and causes of the wealth of nations. Eksmo, 1056 pp
2. Ricardo D (2016) On the principles of political economy and taxation. Eksmo, 1040 pp
3. Mill JS (1981) Principles of political economy. Progress, 448 pp
4. Marshall A (1993) Principles of economics. In 3 volumes. Progress, vol 2, 312 pp
5. Jevons WS (2005) Money and the mechanism of exchange. Socium, 186 pp
6. Edgeworth FY (2003) Mathematical psychics, and further papers on political economy. P. Newman (ed). Oxford University Press, Oxford, 653 pp
7. Sraffa P (1999) Production of commodities by means of commodities. Unity, 160 pp
8. Chamberlin E (1996) The theory of monopolistic competitio. Economics, 350 pp
9. Robinson J (1986) The economics of imperfect competition. Progress, 472 pp
10. Keynes JM (2015) The general theory of employment, interest, and money. Helios ARV, 352 pp
11. Arrow K (1993) The opportunities and limits of a market as a resource allocation mechanism. THESIS: theory and history of economic and social institutions and systems, vol 1, issue 2, pp 53–68
12. Schumpeter JA (2007) the theory of economic development. capitalism, socialism and democracy. Eksmo, 864 pp
13. Hayek F (1989) Competition as a discovery procedure. World Economy and International Relations, vol 12, pp 6–14
14. Simon HA (1993) Rationality as process and as product of a thought. THESIS: theory and history of economic and social institutions and systems, vol 1, issue 3, pp 16–38
15. Stigler GJ (1995) Perfect competition, historically contemplated. a theory of oligopoly. In: Galperin VM (eds) The theory of the firm. SPb.: Economic school, 532 pp
16. Kirzner I (2010) Competition and entrepreneurship. Socium, 288 pp
17. Coase R (2007) The firm, the market and the law: a collection of papers. Novoye Izdatelstvo (New Publishing House), 224 pp
18. Porter M (2016) Competitive advantage: creating and sustaining superior performance. Alpina Publisher, 716 pp

Managing the Manufacturing System Sustainability with the Use of Dissymmetry Approach

Aleksander V. Semenov and Maria Ya. Parfenova

Abstract The issues of application of the dissymmetry approach to analyze the state of the production system and determine the direction of its strategic development are considered in the paper. The dissymmetry approach is based on the separation in the structure of the production system of the static part, which sets the sustainability conditions, and the dynamic part, which ensures the development of the system. The sum vector of interacting asymmetric static and dynamic parts is aimed at achieving the level of their transient equilibrium that meets the requirements of sustainability and competitiveness of production. A mathematical model is proposed to support the adoption of strategic management decisions based on a combination of the approach of dissymmetry and entropy of fuzzy sets. This model allows us to determine the critical point of the phase transition of the production system from one qualitative state to another qualitative state. The critical point of the phase transition indicates the need for changes: increasing the potential of the dynamic part in accordance with modern requirements or increasing the static part or the need to move to a higher system level of production.

Keywords Dissymmetry approach · Entropy of fuzzy sets · Level of transient equilibrium · Point of phase transition

1 Introduction

In the context of the new paradigm of sustainable development of society and the rapid flow of innovations, it is necessary to improve approaches to the management of production systems in order to ensure their effective functioning and compliance with actual requirements. The functioning of any system is characterized by such

A. V. Semenov · M. Ya. Parfenova (✉)
Moscow Witte University, 12, 2nd Kozhukhovsky pas., Bldg. 1, Moscow 115432, Russia
e-mail: mparfenova@muiv.ru

A. V. Semenov
e-mail: asemenov@muiv.ru

A. V. Semenov and I. A. Sokolov (eds.), *Sustainable Development: Society, Ecology, Economy*, Earth and Environmental Sciences Library,
https://doi.org/10.1007/978-3-030-73110-6_9

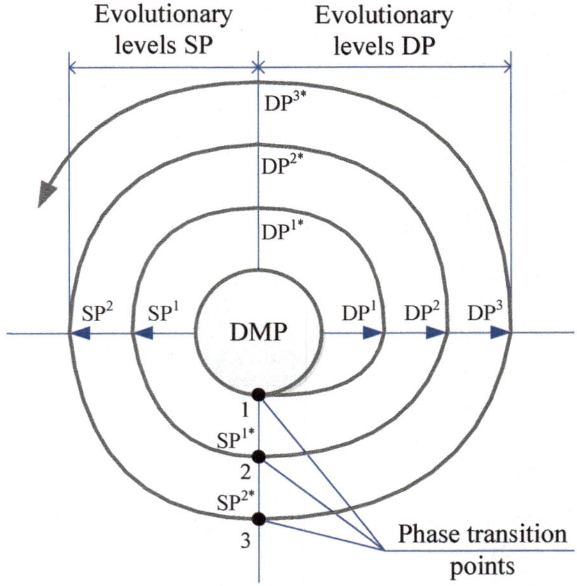

Fig. 1 Diagram of the development of the production system in the framework of the concept of dissymmetry

key indicators as sustainability and efficiency. In this study, the sustainability of the production system is considered as its ability to maintain the specified parameters under conditions of internal and external disturbing impacts. Efficiency is the property of a system to fulfill its aim under specified conditions of use and with a certain quality (Fig. 1).

The controlled object of the production system is considered in the form of interacting static and dynamic parts [1]. In this work, the static part corresponds to the technological process, the dynamic part corresponds to the intellectual process (control process), which are determined by the target functions set for these processes. The static and dynamic parts are considered as asymmetric structures (static and dynamic structures). It is obvious that the belonging of individual components of the production system to a static or dynamic part is not invariable in time and feature space, however, this circumstance does not reduce the practical value of the management principles using these concepts.

The static structure ensures the sustainability of the production system, taking into account the deterministic time. The dynamic structure is aimed at developing and achieving a level of competitiveness. The interaction of static and dynamic structures in a dialectical unity determines the efficiency of the production system. The separation of static and dynamic parts in the production system creates the basis for the analysis of casual effects and the identification of hidden patterns at the system level. The approach to managing the sustainability of the production system based on the dissymmetry of interacting structures is aimed at achieving the level of their transient equilibrium [1] and the evolutionary development of the system.

2 Methodology and Logic of Research

Depending on the magnitude of the arising deviations in accordance with the established threshold value of the parameters in the current state of production processes, the following classification of emerging situations is applied: standard situations are classified as normal if the values of the parameters correspond to the reference (calculated) values; abnormal critical situations, if it is possible to bring them to normal and eliminate losses by a management decision; abnormal emergencies or uncorrectable situations - losses are inevitable. In emergency situations, specialists use their vast experience and knowledge to eliminate them. In this case, the human factor is critical. In accordance with the accepted classification of emerging situations, the following components of the dynamic part and integral indicators of its effectiveness in production management are determined: the quality of a specialist's thinking as the ability to recognize and assess the situation within a reasonable time; the human factor as the probability of making and implementing effective management decisions in a given time interval; human potential as an opportunity to organize competitive production.

In current situations, an analysis of the impact of interacting static and dynamic structures on the course of production and an assessment of the degree of mismatching of their potentials is made. Under the mismatching of the potential levels, the discrepancy of indicators characterizing the state of static and dynamic structures (technological and intellectual processes) is considered. Depending on the level of mismatch in the state of asymmetric structures, control actions are formed aimed at achieving a given level of state in the static part to ensure the sustainability of the production system, or at changing the target function of the asymmetric dynamic part for the development of the production system while maintaining sustainability within specified limits.

Making decisions on the development of static and dynamic structures is accompanied by the procedure for assessing the mismatch of their potential levels and the search for the level of transient equilibrium. The levels of transient equilibrium of the potentials of interacting asymmetric structures in the reference model of production are determined by their equal ratio. The achieved level of mobile equilibrium of potentials is considered as an indicator of the sustainability of production. The mismatch of the potential levels of the static and dynamic structures can lead to a qualitative change in the state of the production system and the need to solve a strategic problem associated with increasing the sustainability and competitiveness of production. Therefore, the level of the transient equilibrium of the potentials of interacting structures shows the need and possibility of developing a particular structure or transferring the production system to a new qualitative level of functioning.

Complex production systems, which are characterized by a multitude of heterogeneous parameters of the controlling object (production process), objectively include non-stationary objects. The parameters of non-stationary objects vary not only in the feature space, but also in time, which causes ambiguity and uncertainty in the recognition and assessment of emerging situations. Certain aspects of non-stationarity of production processes are also present in the functional properties of objects, which are formalized in the control model as stationary due to the limitations of the feature space and the objective incompleteness of information. Even in the presence of a formalized reference model of the controlling object, there is a feature space in which the property of incomplete determinateness manifests itself. In the current situation, it is necessary to make decisions taking into account the accepted restrictions on the values of the parameters of the controlling object, the time of forming and implementation of management actions.

Classical control models consider the nonstationarity factor as a random variable, while it is due to uncertainty factors in time [2]. For this class of production systems, it is more correct to use the requirements of a quasi-real-time system, which ensures the determinateness of the response time given by the interval $[t_{min}, t_{max}]$ to a non-deterministic flow of external events, and also allows a dynamic change in the response time at subsequent control steps, in keeping with the general requirements of determinateness [3]. In this case, the use of quasi-real-time mechanisms that make it possible to model the points of phase transition of the production system from one qualitative state to another qualitative state and form management decisions taking into account the entropy of emerging situations is efficient in the production management model.

In order to create a production process management model, the issues of constructing classification feature of a set of potentially possible situations S, constructing decision functions in situations S are solved. In order to create the tenancy region of the situation S_i in the X feature space, it is necessary to determine the boundaries between the tenancy regions to the situations S_i and S_j, to determine the time of transition of the system from the state S_i to S_j. During the creation of a management model all calculations are carried out on the flow production model, and not on real data. This makes it possible to fulfill the necessary conditions for the applicability of quasi-resonant control mechanisms: to determine a fuzzy set of deviations of the current parameters of production processes from the planned values taken as reference. The use of fuzzy set mechanisms is preferable to statistical methods due to the objective impossibility of matching the uncertainty of the production process parameters to the criteria of the "general set of events" in statistics, as well as due to the relatively simple formalization of linguistic meanings using fuzzy logic mechanisms [4].

There are several known options for postulating the entropy of a fuzzy set, which are an extension of the concept of entropy introduced by Shannon K. In this paper, we use the definition of entropy in accordance with the probability measure of a fuzzy set according to Kolmogorov's axiomatic of probability [5]. In the procedure for creating of a management model, searching of all admissible sequences of execution of production operations and simulation modelling of a system behavior under the impact of a fuzzy set of deviations are performed. This set has a membership function that simulates the Poisson point process. For a fuzzy set of input parameters with entropy $w(t)$, transferring the production system from one state to another, one can compose a system of Kolmogorov ordinary differential equations with variable coefficients in the form of [5]

$$\frac{dP_i(t)}{dt} = \sum_{j=1}^{n} P_j(t)w_{ij}(t) - P_i(t)\sum_{j=1}^{n} w_{ij}(t), \quad i = 1, 2, \ldots, n$$

For any moment of time *t*, the normalization condition $\sum_{i=1}^{n} P_i(t) = 1, \ t \geq 0$ is executed. This follows from the fact that at any moment *t* of the event $\{S(t) = S_1\}, \ \{S(t) = S_2\}, \ldots \{S(t) = S_n\}$ form a complete group of incompatible events [6]. Kolmogorov differential equations for state probabilities $P_1(t)$ and $P_2(t)$ are the following

$$dP_1(t)/dt = P_2(t)\mu(t) - P_1(t)w(t)$$
$$dP_2(t)/dt = P_1(t)w(t) - P_2(t)\mu(t)$$

After transformations, we obtain a linear differential equation with variable coefficients

$$dP_1(t)/dt + [w(t) + \mu(t)]P_1(t) = \mu(t)$$

Taking $\mu(t) = \mu = const, \ w(t) = w = const$, and at the initial condition of $P_1(0) = 1$ the differential equation will be written as

$$dP_1(t)/dt + (w + \mu)P_1(t) = \mu$$

The solution to the last equation will be the entropy of a fuzzy set of system states in accordance with the reference indicators at time *t*

$$P_1(t) = \frac{\mu}{w + \mu} + \frac{w}{w + \mu} e^{-(w+\mu)t}$$

The membership function of a fuzzy set of system states in the presence of deviations at time *t* is equal to

$$P_2(t) = 1 - P_1(t) = \frac{w}{w + \mu}\left(1 - e^{-(w+\mu)t}\right)$$

An entropy approach is used to find the critical point of the phase transition of a fuzzy set of model output values. Let us consider the basic properties of entropy in relation to the production system. The entropy *H(S)* of a source *S* with two states S_1 and S_2 varies from zero to one, reaching a maximum when their probabilities are equal, i.e.

$$P(S_1) = P(S_2) = 0.5, \quad H(S) = -\left[P \log P + (1 - P)\log(1 - P)\right]$$

In conditions of $P << (1 - P)$ the partial uncertainty, for example, per state S_1, is high, however such source states are rarely occurred. In this case, the state S_2 is oftenly realized, however the uncertainty attributable to such a state is very low and the entropy characterizing the average uncertainty per state of the ensemble is also low. A similar situation is observed for $P >> (1 - P)$. Entropy continuously depends on the distribution function of fuzzy state functions, which directly follows from the continuity of the function $P \log P$. In conditions of $P_1(t) = 0.5$ the entropy $H(S) = 1$. At entropy. Let us solve the equation with entropy equal to one, denoting $w + \mu = a$, we will obtain $t = -1/a \ln((0.5a - \mu)/w)$. Substituting the value a, we find the time t

$$t = -\frac{1}{w + \mu} \ln \frac{w - \mu}{2w}$$

The choice of time unit is determined by the specifics of the production process.

The moment of time t is considered as the critical point of the phase transition of the production system from one qualitative state to another qualitative state. Precisely in a system with classes of states S_1 and S_2, a necessary condition for the implementation of quasi-real time mechanisms is satisfied.

The result of production situations modeling are: a lot of classification features X for each situation from the set S; many controlling actions necessary to transfer the system from the state S_i into S_j; the time of the corresponding transition. The data obtained are the basis for making decisions on increasing the potential of static or dynamic structures or the transition of the production system to a new level of functioning.

The cyclical and evolutionary nature of the development of the production system is shown in the figure. When the system moves to a new level, the state corresponding to the standard situation is taken as the initial conditions, a point of transient equilibrium has been reached between the states of static and dynamic structures. The diagram in the figure is interpreted as follows: line 1, DP^1 reflects the increase in the level of the dynamic part in accordance with actual requirements; line DP^1, $DP^{1}*$ reflects the stage of stabilization of the dynamic part; line $DP^{1}*$, SP^1 - increase in the level of the static part; SP^1, $SP^{1}*$ - stage of stabilization of the static part and achievement of sustainability of the production system at point 2. The next stages of the system's development are cyclically repeated at a higher level in regard to the previous level.

The dissymmetry approach makes it possible to determine the area of possible strategic management decisions for a targeted change in the state of static and dynamic structures and the evolutionary development of the production system.

3 Conclusion

The proposed approach to managing the sustainability of a production system based on the dissymmetry of interacting static and dynamic structures makes it possible to reduce non-production costs during the operation of the system, as well as during its transition to a new level of development. This is achieved by determining the critical point of the phase transition of the production system from one qualitative state to another qualitative state. This approach can be transformed to various types of systems and controlling objects.

The proposed mathematical model for supporting the adoption of strategic management decisions using the entropy of fuzzy sets creates the basis for constructing the space of possible and feasible decisions.

References

1. Parfenova MYa, Semenov AV (2015) The model of production system management. Patent of Russia. No. 153307. 2015. Bull. No. 19
2. Dubois D, Prade A (1990) Possibility theory. applications for knowledge representation in informatics, in French. Radio Commun, 288 pp
3. Parfenova MYa, Kolesnikov AA, Sarajev VI (2010) A model of quasi-resonant management of an unsteady object using entropy of a fuzzy set. In: Proceedings of the Institute of System Analysis of the Russian Academy of Sciences: Dynamics of Inhomogeneous Systems, vol 53(3) Issue 14. LENAND, Moscow, pp 29–33
4. Yarushkina NG (2004) Fundamentals of the theory of fuzzy and hybrid systems: textbook for universities. N.G. Yarushkina. Finance and Statistics, Moscow
5. Kolmogorov AN (1987) The theory of information and the theory of algorithms. Nauka, Moscow
6. Ventzel ES, Ovcharov LA (1973) Probability theory. Nauka, Moscow

Cyclical Economy and Innovations

Innovations in Digital Platforms: Project Scripts for Forming a Unified Information Environment

Alexander A. Zatsarinnyy and Alexander P. Shabanov

Abstract The research refers to the scientific problem of creating a unified information environment in the ecosystem of an interdisciplinary digital platform with an aim of modeling new production technologies in the field of emergency response. The object of the research is the digital platform ecosystem configuration management database. The scientific problem of the research presented in the paper is to find a methodological approach to updating the configuration management database through the use of information about a new economic entity—an organizational system of production, scientific, educational, financial, innovative or other profile, introduced into the ecosystem of a digital platform. A method is proposed for forming a project script in the configuration management database for assessing the effectiveness of its functioning as part of an ecosystem. The method is developed on the basis of the results of the analysis of world experience in the application of concepts for IT service management, data processing and transformation—"Information Centric Network", data transmission and processing—"Intent-Based Networking". The scientific novelty of the method is the provision of automatic formation of new records in the configuration management database, taking into account the records already available in the database about configuration elements that affect the functioning of the new organizational system. The practical significance of the study lies in reducing the time frame for updating the structure of the configuration management database in the segment of assessing the effectiveness of organizational systems in the ecosystem of an interdisciplinary digital platform.

Keywords Organizational system · Functioning · Efficiency · Digital platform · Unified information environment · Project script · Innovation

---

A. A. Zatsarinnyy · A. P. Shabanov (✉)
Federal Research Center "Computer Science and Control, Russian Academy of Sciences" of the, Vavilova st. 44-2, Moscow 119333, Russia
e-mail: apshabanov@mail.ru

A. A. Zatsarinnyy
e-mail: alex250451@mail.ru

A. V. Semenov and I. A. Sokolov (eds.), *Sustainable Development: Society, Ecology, Economy*, Earth and Environmental Sciences Library,
https://doi.org/10.1007/978-3-030-73110-6_10

1 Terms of Reference

Ecosystems of the digital economy of the Russian Federation are characterized by the accumulation and transformation of information in the form of digital data. Such ecosystems are a key factor of production in all areas of socio-economic activity. In ecosystems based on digital platforms, effective interaction is carried out, including cross-border spillover, interaction between business, interaction within the scientific and educational community, and interaction between the state and citizens. Platforms and technologies of the digital economy provide the formation of competencies for the development of markets and sectors of the economy (spheres of activity) [1]. Unified information environments in such platforms cover regulatory processes, information infrastructure and others reproduced in ecosystems of various technological orientations—big data, neurotechnology and artificial intelligence, distributed ledger systems, quantum technologies, new production technologies, the industrial Internet, robotics and sensorics components, wireless communication technologies, technologies of virtual and augmented reality [2–10].

The study refers to the *scientific problem* of creating and evolutionary development of a unified information environment in the ecosystem of an interdisciplinary digital platform with and aim of modeling new production technologies. Advanced technologies create new markets and industries, contribute to the growth of labor productivity, increase the competitiveness of individual sectors and national economies as a whole [11]. Within the framework of this problem, scientific problems are being solved, the formulation of which is conditioned by the implementation of national projects in the spheres of management and sectors of the economy. When studying this problem, an analysis of research results related to various types of activities and areas of application was carried out, including resource management of high-performance clusters [12], computations in solving problems of processing data streams [13] and geospatial data [14], emergency response [15] and the development of a distributed linguistic decision support system [16], the training of neural network cascade classifiers [17] and the analysis of trends in changing the principles of enterprise management in the context of the digital economy development [18]. A specific property of the research carried out is the reproduction of the processes of collection and accumulation of initial data on all entities that affect the subject of research, and innovative processes of data conversion to achieve target settings. The initial and derived data from them in their totality and the relationship between themselves and with the external environment form a single information environment.

Solving the *problem* of developing a unified information environment is of particular relevance in the areas of management and security. In this regard, the object of the study is a configuration management database in the ecosystem of an interdisciplinary digital platform with an aim of modeling new production technologies in the field of emergency response.

The *scientific problem* of this study is to find a methodological approach to updating the configuration management database through the use of information about a *new economic entity—the organizational system* of production, scientific, educational, financial, innovative or other profile, introduced into the ecosystem of the digital platform. The content of the configuration management database is associated with the digital platform ecosystem's unified information environment model.

The solution to the problem was carried out on the basis of the following:

- on the purpose and functionality of situational centers, taking into account the provision of integral information support for management decisions in a network environment, the development of technologies for collective and artificial intelligence, cognitive approach, expert procedures, convergent management [19];
- on the awareness of the importance of connections: on the one hand, between the degree of effectiveness of situational centers and the degree of effectiveness of the functioning of organizational systems, to support which these centers function; on the other hand, between the degree of efficiency of organizational systems and factors influencing their activities, moreover, the impact of factors on the degree of efficiency is manifested through the display in the databases of configuration units of situational centers of indicators of states of objects of observation in a controlled space [20];
- about the general approach and initial data for the projection of information and telecommunication networks, the main stages of projection, their content and features [21];
- on the main processes of situational centers—monitoring the state of the control object, predicting the development of the situation, modeling the consequences of management decisions and their optimization based on the configuration management database (configuration units database) [22];
- on the approach to information support for predicting scientific and technological achievements, which: provides collection of information on intellectual property items related to critical technologies and published in various sources of leading world states; assesses the effectiveness of domestic and foreign technologies in comparison with world standards and develops the predictions—scripts of decisions on the development of technologies [23].

Situational centers are associated with hardware and software systems in the ecosystem of the digital platform that reproduce similar processes.

A model of the configuration management database segment of the ecosystem of a digital platform purposed for the assessment of the performance of organizational systems introduced into this ecosystem is considered in the paper. An overview and constructive analysis of innovative methods and models to support the activities of organizational systems is provided.

Based on the results of the analysis, a method was developed for forming a project script in the configuration management database to assess the performance of a new organizational system introduced into the ecosystem of a digital platform. The sequence of steps in the method is structured taking into account the world experience in building processes for collecting, processing, displaying and archiving information, processing and transforming data in computer networks, transferring and processing data in data transmission networks [24–29] and Russian experience in solving problems of creating new businesses-models [30] and their transformation [31], digital transport with global navigation satellite systems [32], strategic [33, 34] and fault-tolerant [35] technological platforms, for knowledge management [36] and assessing the level of security of computer systems [37].

The creation of a unified information environment in the configuration management database in combination with new production technologies—innovative methods and models to support the activities of organizational systems, allows providing the automated involvement of new organizational systems in the digital platform ecosystem with the ability to control the effectiveness of their activities.

2 Configuration Management Database

The concept of IT Service Management [38] is proposed as the main methodological tool for studying the issues of structuring the configuration management database and its role in the processes of assessing the effectiveness of organizational systems. The following terms and definitions related to this concept are used in the paper [41]:

- a Configuration Management Database (CMDB) is a database used to store configuration records throughout their life cycle. The configuration management system maintains one or more configuration management databases, and each database stores attributes of configuration items and links to other configuration items;
- a Configuration Management System (CMS) is a collection of tools, data, and information used to support the management of service assets and configurations. The CMS is part of the overall knowledge management system of services and includes tools for collecting, storing, managing, updating, analyzing and presenting data about all configuration items and their relationships. The CMS may also contain information about incidents, problems, known bugs, changes and releases. The CMS is supported by Service Asset and Configuration Management and is used by all IT Service Management processes;
- a configuration record is a record that contains information about a configuration item. Each configuration record documents the life cycle of an individual configuration item. Configuration records are stored in the configuration management database and maintained as part of the configuration management system;
- a configuration item (CI) is any component or other service asset that needs to be managed to deliver an IT service. Information about each configuration item is recorded as a configuration record in the configuration management system. This record is maintained throughout its life cycle through service asset and configuration management processes. Configuration items are under the control of change management. These usually include IT services, hardware, software, buildings, people, and white papers such as technology documents and service level agreements (SLAs);

- IT service is based on the use of information technology and provides support for business processes. To perform an IT service, roles, processes, technologies are assigned and must be defined in a service level agreement.

The application of the IT Service Management concept ensures the filling of the configuration management database with components relevant to new (innovative) production technologies, including the processes that ensure the provision of the IT service "Evaluation of the performance of organizational systems from the digital platform ecosystem." Thus:

- each configuration element, regardless of its complexity—is it an organizational system or a personal computer, has its own configuration record in the database;
- each innovation process or group of processes, including processes for assessing the effectiveness of functioning, has its own group of configuration records, interconnected with each other and with other groups (processes);
- such groups are electronic models of the relevant innovation processes and are positioned in the database as its segments, for example, a segment for assessing the effectiveness of the digital platform ecosystem;
- During the reproduction of the functions of innovative processes, configuration records are copied and operations of processing, transformation, transfer, etc. are performed on them, which are due to the logic of the execution of these functions, each configuration record about the process, hardware and software product, attribute, communication and other entities that affect the effectiveness of organizational systems, and a group of configuration records can be represented as a project script;
- the composition of configuration records in database segments, including in the segment for assessing the effectiveness of the ecosystem of the digital platform, are subject to changes in accordance with changes in the information infrastructures of organizational systems;
- access to the configuration management database and to work with configuration records is provided to the processes of configuration management, changes, incidents, problems and others, in accordance with the rules established for them.

The configuration management database is located in the data center of the digital platform, the information model of which is shown in Fig. 1.

The digital platform in accordance with the model (Fig. 1) is built on the basis of technologies of information, control, navigation systems, which determine the main directions of scientific and technological development [39], and which are used to support the activities of organizational systems with ensuring the maximum degree of automation of management processes on based on the preparation and application of a priori and retrospective scripts for their automatic execution by robotic objects, information systems or for decision-making by control subjects, depending on the available resources and the prevailing circumstances in the environment of activity [15, 40].

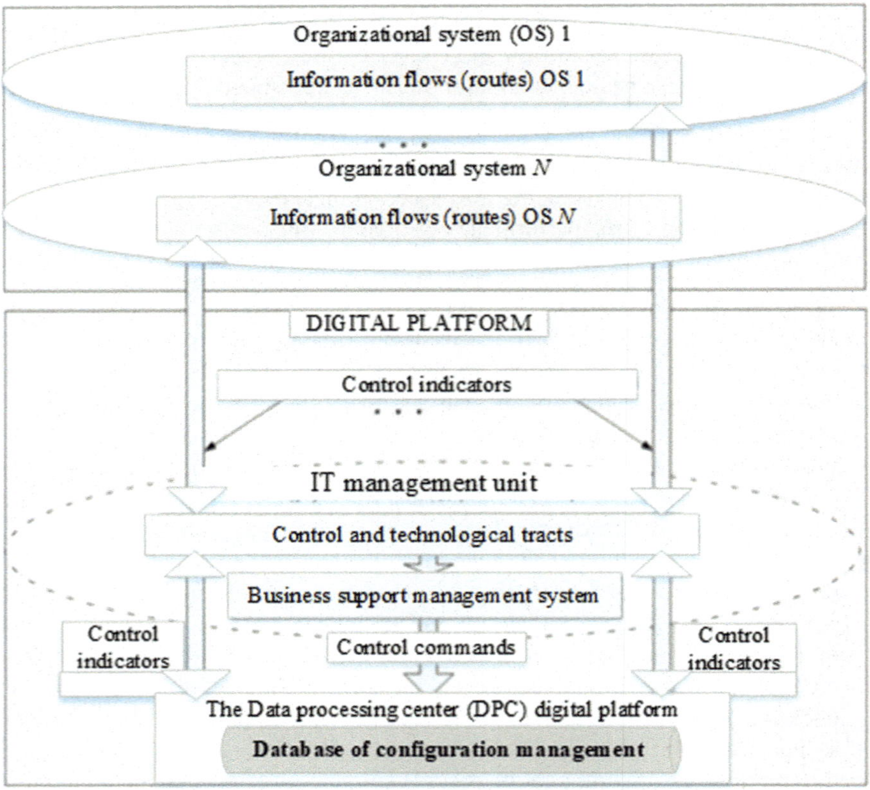

Fig. 1 Model of the digital platform

3 Analysis of Innovative Methods and Models of the Support of Organizational Systems Functions

The main components of digital platforms and organizational systems consolidated on their basis are information systems. ***Due to this fact***, it is obvious that the existing methodological, scientific, technical and other groundwork created for the projecting of information systems will also be in demand in the projecting of digital platforms that include a database of configuration elements. ***Taking this factor into account***, we analyzed a sample of new production technologies—innovative methods and models for ensuring information interaction between organizational systems in the ecosystem of a digital platform. Their characteristic project scripts are displayed in Table 1.

Another example of new production technologies is the method and models in the field of supporting the activities of organizational systems. Their characteristic project scripts are shown in Table 2.

Table 1 Innovative management method and models (example 1)

Method, model	Characteristic
1. Integrated control system (https://www1.fips.ru/ofpstorage/Doc/IZPM/RUNWC1/000/000/002/630/393/ИЗ-02630393-00001/document.pdf)	Information interaction is provided through a project script for converting data about the control command, generated in the codes of one system, into data about the same command, however in the codes of another system
2. Method of transmission of control commands (https://www1.fips.ru/ofpstorage/Doc/IZPM/RUNWC1/000/000/002/631/147/ИЗ-02631147-00001/document.pdf)	Information interaction is provided through the project script for converting data about the control command into data about its function and reverse transformation on the receiving side, however in different codes
3. Method of information transmission (https://www1.fips.ru/ofpstorage/Doc/IZPM/RUNWC1/000/000/002/618/366/ИЗ-02618366-00001/document.pdf)	Information interaction is provided on the transmitting side due to the project script for converting data about information into data on its purpose, on the receiving side and due to the reverse transformation, however in different codes
4. Control device (https://www1.fips.ru/ofpstorage/Doc/IZPM/RUNWU1/000/000/000/167/247/ПМ-00167247-00001/document.pdf)	Information interaction is provided through the project script for the remote control of objects, regardless of the correspondence of the data codes on the control command and the data codes in the object
5. Complex of information interaction (https://new.fips.ru/Archive4/PAT/2016FULL/2016.03.10/DOC/RUNWU1/000/000/000/160/257/DOCUMENT.PDF)	This complex implements the project script of method of information transmission (paragraph 3)
6. Data transmitter (https://new.fips.ru/Archive4/PAT/2016FULL/2016.11.10/DOC/RUNWU1/000/000/000/165/924/DOCUMENT.PDF).	This transmitter implements the project script of method of transmission of control commands (paragraph 2) on the transmitting side
7. Data receiver (https://new.fips.ru/Archive4/PAT/2016FULL/2016.11.10/DOC/RUNWU1/000/000/000/165/993/DOCUMENT.PDF)	This receiver implements the project script of method of transmission of control commands (item 2) on the receiving side

Based on the results of the analysis of new production technologies—methods and models, it can be noted that:

- the sources of data used in the corresponding processes of collection, processing, transformation and transmission-reception are configuration management databases or their analogs;
- the project scripts used to execute the processes are configuration items, the descriptions of which are presented in the configuration management databases in the form of configuration records;

Table 2 Innovative management methods and models (example 2)

Method, model	Characteristic
1. Method of supporting operation of organizational system (https://new.fips.ru/Archive/PAT/2014FULL/2014.11.10/DOC/RUNWC2/000/000/002/532/723/DOCUMENT.PDF)	Provides an increase in the efficiency of support for the activities of organizational systems due to the automatic execution of project scripts for assessing the indicators of configuration elements and automatic control, taking into account the performed assessment
2. System of situationally analytical centers of organizational system (https://new.fips.ru/Archive/PAT/2014FULL/2014.11.20/DOC/RUNWC2/000/000/002/533/090/DOCUMENT.PDF)	The increased efficiency of the decision support process is achieved due to the automated development of project scripts for solving problem situations in a distributed system of situational centers
3. Organizational systems management system (https://new.fips.ru/Archive4/PAT/2016FULL/2016.08.27/DOC/RUNWC1/000/000/002/595/335/DOCUMENT.PDF)	The increased reliability of project management scripts is reached while preventing threats to organizational systems of different hierarchy levels
4. Management Center of Organizational system (https://new.fips.ru/Archive/PAT/2013FULL/2013.04.27/DOC/RUNWU1/000/000/000/127/493/DOCUMENT.PDF)	The reduction of the number of problem situations and reduction of the time for their resolution, if they have already arisen, is provided due to the centralized development of project scripts for solving problem situations
5. Monitoring Center for sustainability information systems (https://new.fips.ru/Archive/PAT/2013FULL/2013.07.10/DOC/RUNWU1/000/000/000/130/109/DOCUMENT.PDF)	Expansion of the functionality of management due to the ability to automatically generate, save, display and transfer of the project scripts with a description of incidents for the intended purpose
6. Stability of information systems support center (https://new.fips.ru/Archive/PAT/2013FULL/2013.09.10/DOC/RUNWU1/000/000/000/132/227/DOCUMENT.PDF)	Expansion of the functionality of management, due to the automatic generation, storage, display and transfer of the project scripts with a description of the problems for the intended purpose

As it can be seen from the characteristics of methods and models presented above (Tables 1 and 2), the processes for developing project scripts are based on the information obtained as a result of actions for the collection, processing and analysis of initial data characterizing the entities of the corresponding subject area. The individuality of the project scripts is determined by the semantic content of this information.

Based on the above, a statement was formulated about the prospects of developing a method for generating data on a project scenario based on a digital platform for a subject area—a unified information environment for a conglomeration of organizational systems.

4 Formal Model of the Working Efficiency Assessment Process

Based on the results of the analysis of the characteristics of new production technologies—innovative methods and models for supporting the activities of organizational systems from the ecosystem of the digital platform, and also taking into account the assumption of the representation of configuration records as project scripts, a formal model was developed to describe the corresponding segment of the management database configuration—by assessing the working efficiency.

The working efficiency assessment model in the configuration management database has the following levels.

The first level is systemic, in which

$$S = \{S_1, S_2 \ldots S_N\} \quad (1)$$

is a set of configuration records about configuration items S_1, S_2 … S_N, representing organizational systems in the digital platform ecosystem, where N number of organizational systems.

With each of the subsets S_1, S_2 … S_N configuration elements of the second level are correlated—objects of the internal information infrastructure and the external environment—data transmission networks, engineering structures, etc. When configuration records for these items are placed in the configuration management database, they are converted to eliminate duplicate records. As a rule, these are records about items of the external environment.

The second level is an object level, in which

$$V = \{V_1, V_2 \ldots V_N\} \quad (2)$$

is a set of configuration records about configuration items V_1, V_2 … V_N, representing subsets of objects in the ecosystem of the digital platform that are used to assess the performance of organizational systems, respectively S_1, S_2 … S_N.

With each of the subsets V_1, V_2 … V_N the configuration elements of the third level are correlated—indicators of the state of these objects of the internal information infrastructure and the external environment.

The third level is indicator, in which:

$$V_i = \{V_{11}, V_{12} \ldots V_{1Mi}\} \quad (3)$$

are subsets of configuration records about configuration items, which are *normalized* indicators of the state of the corresponding objects in S_i organizational system;

$$V_i^* = \left\{V_{11}^*, V_{12}^* \ldots V_{1Mi}^*\right\} \quad (4)$$

are subsets of configuration records about configuration items, which are *actual* indicators of the state of the corresponding objects in S_i organizational system, where Mi is a number of indicators, $i = 1, 2\ldots N$.

Normalized indicators of configuration items are determined when preparing an organizational system for its introduction into the ecosystem of a digital platform. The normalized indicators are recorded in the agreement concluded between the relevant business entity and the enterprise providing it with the IT service "Assessment of the working effectiveness of the organizational system."

The actual indicators of configuration items are determined at the stage of operation of the digital platform, for example, using the new production technology "Monitoring Center for sustainability information systems" (https://new.fips.ru/Archive/PAT/2013FULL/2013.07.10/DOC/RUNWU1/000/000/000/130/109/DOCUMENT.PDF), which monitors the state of configuration items in information systems and records their performance in real time.

At the stage of operation of the digital platform, in accordance with the established regulations, the effectiveness of each organizational system and, in general, the ecosystem of the digital platform is assessed. The solution to this problem is carried out with the help of new production technologies. An example of such technology is "Method for determining innovative objects in information systems" [5, p. 116]. Using this method, the indicators of the configuration elements of the ecosystem of the digital platform are investigated—servers, computing complexes, devices and storage systems, devices of the local computer network and the backbone data transmission network and other hardware and software tools that perform functions (operations) in accordance with user requests on the processing and transmission of data with an impact on the order of their arrangement and their content.

When introducing a new digital platform into the ecosystem S_{N+1} of the organizational system, the configuration management database should be expanded by placing in it configuration records about the configuration elements of this organizational system, subject to ensuring that its performance is assessed. This property of the model is fundamentally important for carrying out operations to update the ecosystem of the digital platform when a new economic entity is introduced into it.

5 Method of Generating Data on the Project Script for Assessing the Effectiveness of the Organizational System

Taking into account the factor of writing to the configuration management database data on configuration items used in the processes of assessing the performance of project script data in this database of the term "project script" as identical to the term "configuration record".

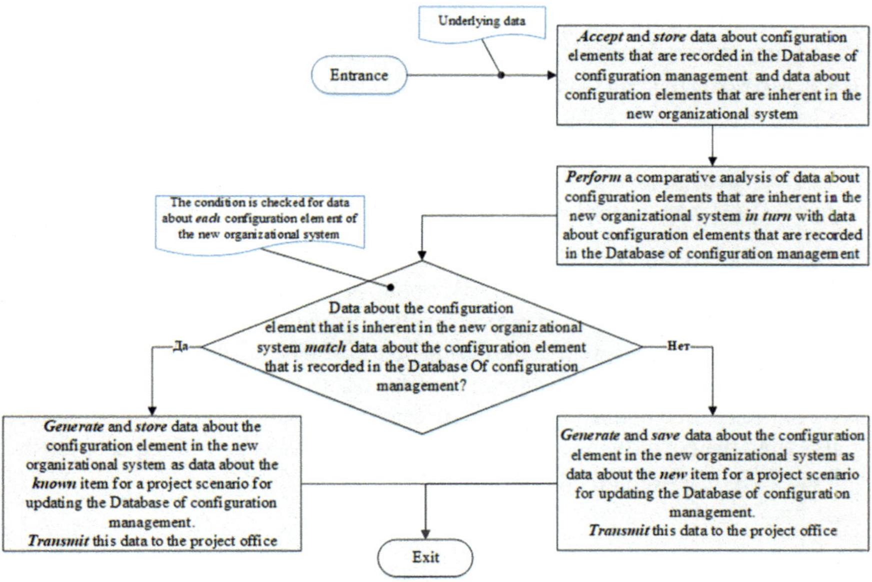

Fig. 2 Scheme of the method for generating data on the project script

A diagram of the method for generating data on the project script is shown on Fig. 2. A feature of this scheme is the refinement of the structure of project scripts to their elements, which is due to the intention to display the complexity of the structure of configuration records in real databases.

The purpose of this method is to reduce the time spent on innovative projects in the segment of assessing the effectiveness of new organizational systems that are introduced into the ecosystem of the digital platform.

The following data are used in the operations of this method:

- project scripts—configuration records that are located in the configuration management database of the digital platform ecosystem;
- project scenarios—configuration records about a new organizational system, in respect of which a decision is made either to place them in the configuration management database or not to place them, if these records are already present in it.

Let us accept additional restrictions:

- configuration records for the new organizational system exist and are placed in the configuration management database in its infrastructure;
- the tools that are necessary for using these records in the project to update the ecosystem of the digital platform in terms of assessing the performance of organizational systems are known and there is a possibility to apply them;

- as working tools in this project, well-known software applications for process automation can be used: Change Management, Configuration Management, Release Management, Configuration Administration, Capacity Management and others, created in accordance with the concept of IT Service Management;
- the executors of this project can be providers of leading world or domestic IT Service Management software vendors, for example, a fairly complete list of them is presented in [38].

The novelty of the method lies in the automatic generation of new records in the configuration management database, taking into account the already existing records of configuration elements that affect the functioning of the new organizational system and, in general, the ecosystem of the digital platform. The practical significance of its applicability is due to the reduction in the time frame for building the competencies and resources of the digital platform ecosystem in the field of ensuring the production of assessing the performance of organizational systems as part of the ecosystem of an interdisciplinary digital platform.

6 Conclusions

The study refers to the *problem* of creating and evolutionary development of common information environments in the ecosystems of the digital economy. The formulation of the task of finding a methodological approach to updating the configuration management database through the use of information about a new business entity introduced into the ecosystem of an interdisciplinary digital platform built using new production technologies is substantiated.

The structure of the configuration management database as part of the digital platform and the concept of information technology service management—IT Service Management, are proposed to study the issues of structuring the configuration management database and its role in the processes of assessing the effectiveness of organizational systems.

The analysis of innovative methods and models to support the activities of organizational systems is carried out taking into account the world and Russian experience in the use of critical information technologies in the field of management and sectors of the economy.

A multi-level formal model of the process of assessing the effectiveness of the activities of organizational systems and, in general, the ecosystem with a designated purpose for its use in the management of information technology services in accordance with the concept of IT Service Management has been developed.

A new method has been developed for generating data on a project scenario to assess the effectiveness of an organizational system in configuration management databases in ecosystems of interdisciplinary digital platforms that are built using new production technologies or aimed to use them. The novelty of this method lies in ensuring the minimization of the number of configuration records with the automatic generation of data on the configuration elements of the organizational system entered into the ecosystem in the ecosystem configuration management database of the digital platform. The practical significance of the method lies in reducing the time spent on project work on introduction of a new participant into the conglomeration of organizational systems that conduct joint activities on the basis of a digital platform.

Ackonowledgements This work is partially supported with grants 18-29-03091 и 18-29-03100 by the Russian Foundation for Fundamental Researches.

References

1. The program "Digital Economy of the Russian Federation." Order of the Government of the Russian Federation of July 28, 2017 No. 1632-p. Official Internet portal of legal information [Electronic resource], 2017. http://static.government.ru/media/files/9gFM4FHj4PsB79I5v7yLVuPgu4bvR7M0.pdf (appeal date 21/08/2020)
2. Ognivtsev SB (2018) The concept of a digital platform for the agro-industrial complex. Int Agric J 2(362):16–22
3. Grigoriev MN, Maksimtsev IA, Uvarov SA (2018) Digital platforms as a resource for increasing the competitiveness of supply chains. J St. Petersburg State Univ Econ 2(110):7–11
4. Maksimtsev IA (2018) Digital platforms and digital finance: problems and development prospects. J St. Petersburg State Univ Econ 1(109):7–9
5. Shabanov AP (2018) Innovative management of digital platforms in the knowledge economy. Control Syst Commun Secur 3:106–135. https://sccs.intelgr.com/archive/2018-3/06-Shabanov.pdf. Accessed 21 Aug 2020
6. Potapov DV (2018) On the creation of a digital platform for interaction during pre-trial proceedings between the supervising prosecutor and the preliminary investigation body. J Proc Acad Manag Ministry Internal Affairs of Russia 1(45):85–89
7. Korovin YaS, Tkachenko MG (2017) Hardware and software platform for building a digital field system. Oil Industry 1:84–87
8. Bolshakov SN, Leskova IV, Bolshakova YuM (2017) Digital economy as a component of the technological platform of state policy and management. Management Issues 1(25):64–70
9. Volkov D (2017) Platform of the digital era. Open systems. DBMS 3:38–41
10. Popkov SP (2017) Digital technologies of the Republic of Belarus: a multiservice platform for all subscribers. Econ Strateg 19(8)(150):100–109
11. Dezhina I, Ponomarev A (2014) Advanced manufacturing: new emphasis in industrial development. Foresight-Russia 8(2):16–29
12. Volovich KI, Denisov SA, Shabanov AP, Malkovsky SI (2019) Aspects of the assessment of the quality of loading hybrid high-performance computing cluster. In: CEUR Workshop Proceedings ITHPC 2019—Short Paper Proceedings of the 5th international conference on information technologies and high-performance computing, pp 7–11
13. Talalaev AA, Fralenko VP (2018) Fault-tolerant system for organizing high-performance computing for solving problems of processing data streams. Program systems: theory and applications, vol 36, no 1, pp 85–108. doi:10.25209 / 2079-3316-2017-9-1-85-108
14. Basargin AA, Bugakov PYu, Katsko SYu (2019) Solution of logistic tasks based on processing geospatial data using the TRACKING ANALYST Module as Part of ARCGIS. Vestnik of the Siberian State University of Geosystems and Technologies (SSUGT), vol 24, № 1, pp 83–96. doi:10.33764/ 2411-1759-2019-24-1-83-96
15. Kozlov SV, Shabanov AP, Kubankov AN (2019) Innovations in control systems of actions of robotic objects in the field of emergency response. 2019 Wave electronics and its application

in information and telecommunication Systems (WECONF 2019), pp 8840139. Doi:10.1109/ WECONF.2019.8840139
16. Demidovsky AV, Babkin EA (2019) Development of a distributed linguistic decision support system. Bus Inform 13(1):18–32. 10.17323/ 1998-0663.2019.1.18.32
17. Teplyakov LM, Gladilin SA, Shvets EA, Nikolaev DP (2018) Training of neural network-based cascade classifiers. Inf Process 18(4):439–450
18. Ivanov DA, Ivanova MA, Sokolov BV (2018) Analysis of transformation trends in enterprise management principles in the era of industry 4.0 Technology. SPIIRAS Proc 60(5):97–127. doi:10.15622/ sp.60.4
19. Zatsarinny AA, Ilyin NI, Kolin KK, Lepsky VE, Malinetskiy GG, Novikov DA, Raikov AN, Silvestrov SN, Slavin BB (2017) Situational development centers in a polysubject environment. Control Sci 5:31–42
20. Zatsarinny AA, Shabanov AP (2013) Systemic aspects of the effectiveness of situational centers. Bull Witte Univ. Series 1: Econ Manag 2(4):110–123
21. Zatsarinny AA, Ionenkov YuS, Kozlov SV (2010) Some issues of projecting of information and telecommunication systems. IPI RAN, 218 pp
22. Zatsarinnyy AA, Shabanov AP (2011) Situational Centers: information—processes—organization. Telecommunications 6:42–46
23. Zatsarinnyy AA, Kozlov SV, Shabanov AP (2015) On informational support of activities in the management systems of critical technologies based on situational centers Control Syst Commun Secur 4:98–113. http://sccs.intelgr.com/archive/2015-04/05-Zatsarinnyy.pdf. Accessed 21 Aug 2020
24. Elhady A, El-bakry H, Elfetouh A (2019) Comprehensive risk identification model for SCADA systems. Secur Commun Netw 3914283 (2019). doi:https://doi.org/10.1155/2019/3914283
25. Yin L, Gao Q, Zhao L, Zhang B, Wang T, Li S, Liu H (2020) A review of machine learning for new generation SMART dispatch in power systems. Eng Appl Artif Intell 88: https://doi.org/ 10.1016/j.engappai.2019.103372
26. Mahdavinejadab M, Rezvan M, Adibi P, Sheathe A, Barekatain M, Barnaghi P (2018) Machine learning for internet of things data analysis: a survey. Digit Commun Netw 4(3):161–175. https://doi.org/10.1016/j.dcan.2017.10.002
27. Jia Z, Wang J, Chen X, Sheng Y, Zheng K (2019) An SDN-based measurement scheme to build delay database for time-sensitive network scheduling. Int J Innov Comput Inf Control 15(4):1271–1286. https://doi.org/10.24507/ijicic.15.04.1271
28. Dos Santos P, Travassos G (2016) Scientific knowledge engineering: a conceptual delineation and overview of the state of the art. Knowl Eng Rev 31(2):167–199. doi:https://doi.org/10. 1017/S0269888916000011
29. Yu H, Zhang B (2019) A hybrid MADM algorithm based on attribute weight and utility value for heterogeneous network selection. J Netw Syst Manag 27(3):756–783. https://doi.org/10. 1017/S0269888916000011
30. Gretchenko AI, Gorokhova IV (2019) Digital platform: a new business model in the russian economy. Vestnik of the Plekhanov Russian University of Economics, vol 103, no 1, pp 62–72
31. Olifirov AV, Makoveichuk KA, Petrenko SA (2019) Transformation of business models in the digital economy. Int J Open Inf Technol 7(4):85–91
32. Sokolov IA, Misharin AS, Kupriyanovskiy VP, Pokusaev ON, Lipuntsov YuP (2019) Digital transport projects with global navigation satellite systems—the road to building integrated digital transport systems. Int J Open Inf Technol 7(1):49–77
33. Filin SA (2019) strategic technological platforms as the basis for technological security of Russia in the future. National Interests: Priorities and Security 15(1) (370), pp 18–34. doi:10.24891/ ni.15.1.18
34. Titov VV, Bezmelnitsyn DA (2019) Optimizing strategic management of high-tech business development based on the industrial cluster platform. Region: Econ Sociol 101(1):250–270. doi:10.15372/ REG20190111
35. Pashkov VN (2019) Fault-tolerance distributed control plane for software defined networks. Model Anal Inf Syst 26(1)(79):101–121. doi:10.18255/ 1818-1015-2019-1-101-121

36. Batov G (2019) Information and digital technologies in the knowledge management system of the region. Probl Theory Pract Manag 2:51–58
37. Grusho AA, Grusho NA, Zabezhaylo MI, Timonina EE (2019) Methods of estimation of security of computer systems of information support of digital economy. Int J Open Inf Technol 7(4):61–66
38. Zimin K (2019) Results of the all-Russian research IT Service Management 2019. Information Management [Electronic resource] http://www.itsmforum.ru/upload/medialibrary/559/5593dd51c871d46d2f6c1362fe3026bf.pdf. Accessed 21 Aug 2020
39. Shestakova IG (2019) New temporality of digital civilization: the future has already come, St. Petersburg State Polytech Univ J. Hum Soc Sci10(2):20–29. https://doi.org/10.18721/jhss.10202
40. Zatsarinnyy A, Shabanov A (2019) Model of a prospective digital platform to consolidate the resources of economic activity in the digital economy. In: Proceedings of the 13th international symposium "Intelligent Systems 2018" (INTELS'18). Procedia computer science, Elsevier B.V, vol 150, pp 552–557. https://doi.org/10.1016/j.procs.2019.02.092
41. ITIL® Glossary of Terms English v.1.0. AXELOS Limited [Electronic resource]. 2011. https://www.axelos.com/corporate/media/files/glossaries/itil_2011_glossary_gb-v1-0.pdf

Sustainable Development: Waste Recycling and Circular Economy

Ivan N. Kamyshnikov, Tamara S. Smirnova, and Alexey I. Tikhonov

Abstract The urgency of the problem associated with the implementation of ensuring the transition to rational models of consumption and production, in particular by significantly reducing the volume of waste by taking measures to prevent their formation, reduction, recycling and reuse is shown in the paper. Attention is being paid to the need to move towards an economy that is resource efficient, low-carbon and climate-resilient. The development of such an economy represents a new global challenge for achieving sustainable and inclusive growth. Projects aimed at the development of a closed-cycle economy and rational waste processing are being considered. It is concluded that there is a need for a paradigm shift that will lead to a new industrial policy aimed at ensuring sustainability and innovation that can force us to rethink the model of consumption and business. The circular economy is one of the main development models for the coming years, reducing environmental impact, combining the needs of sustainable growth and combating waste with the needs of competitiveness and innovation.

Keywords Wasteful consumption and production · Circular economy · Sustainability and growth · Recycling

I. N. Kamyshnikov (✉)
University of Macerata, Via Crescimbeni, 30/32-62100 Macerata, Russia
e-mail: ivankamyshnikov@outlook.com

T. S. Smirnova · A. I. Tikhonov
Moscow Aviation Institute (National Research University), Volokolamskoe Highway, 4, Moscow 125993, Russia
e-mail: tamsmir@mail.ru

A. I. Tikhonov
e-mail: mai512hr@mail.ru

A. V. Semenov and I. A. Sokolov (eds.), *Sustainable Development: Society, Ecology, Economy*, Earth and Environmental Sciences Library,
https://doi.org/10.1007/978-3-030-73110-6_11

1 Introduction

The period after World War II, which was characterized by rapid population growth and better living conditions, a potential dichotomy between economic growth and environmental protection emerged. Many documents, beginning with the MIT and Club of Rome report "The Limits to Growth" in 1972, raised concerns about the current model of economic development based on unlimited growth in the consumption of available resources and capital. Despite the opportunities of modern market systems, increasingly based on transnational trade, innovative financial instruments and globalization, the current model of economic growth threatens the sustainable and stable development of mankind and the economy [1].

The purpose of this study is to consider the model of a circular economy in Europe, as well as the existing experience in the implementation of the task of reducing the volume of waste by taking measures to prevent its generation, reduce it, recycle and reuse it, determine the most efficient consumption and production model and determine the most effective economic model. development.

In the process of research, such general scientific methods as analysis and synthesis, consistency and complexity were used.

2 Discussion of the Problem

In this context, the big challenge that many countries will face is negative environmental and social dynamics to which they must respond effectively while maintaining the competitiveness of the production system. A paradigm shift that will lead to new industrial policies aimed at sustainability and innovation that can force us to rethink our consumption and business dealing patterns needs to be implemented.

The transition to a circular economy requires structural change, and innovation is the cornerstone of this change.

The Agenda for Sustainable Development (the 2030 Agenda for Sustainable Development), adopted at the UN Summit on Sustainable Development, held in New York on 25 September 2015, contains 17 goals (Sustainable Development Goals—SDGs) that should be achieved by 2030, resulting in global sustainable development.

The ultimate goal of this challenge is to engage communities around the world in laying the foundations for a better life on the planet and its inhabitants.

If we look at the production of goods worldwide, 60% of its waste ends up in landfills or incinerators, only 40% is reused/recycled and (mostly) only 5% of the cost of the feedstock is recovered [2]. When we turn an asset into waste, we interrupt its information chain that can return a product to the market as a new raw material or semi-finished product. Once a product has become a waste, we can intervene in the processing of its components, but mainly at the "macro level" (wood, plastic, paper, etc.).

In a linear economy, waste generation is very high and increases with consumption (assuming that resources are abundant, available and can be disposed of at a low cost). In a recycling economy, we have efficient cycles that allow us to recycle or reuse products before sending them to landfills, thereby reducing waste, but not eliminating the problem.

Moving to a resource efficient, low carbon, and climate resilient economy is a new global challenge for achieving sustainable and inclusive growth.

A global population will reach more than 9 billion [3] by 2050, [3] and developing countries will experience rapid economic growth. Therefore, the demand for natural resources, especially raw materials, is expected to continue to grow exponentially over the next decades. This trend will also lead to increased environmental and climate impacts unless policies and measures are taken to use resources more efficiently.

In this context, the spread of a new "circular" model of production and consumption is an element of strategic importance for achieving global sustainability goals and at the same time is a factor for improving the competitiveness of countries [4].

At the international level, a broader concept of resource efficiency has been developed in recent years through numerous initiatives in areas such as the OECD, the UNEP International Panel of Experts (UNEP-IRP) and the G7/G8/G20 [5].

The initiative of the German G7 presidency in 2015 follows the path [6] paved by Japan during the G8 presidency in May 2008, within which the "3R Action Plan—Reduce, Reuse, Recycle" was adopted in Kobe [7], containing a series of actions aimed at increasing resource productivity, promoting the "recycling society" and the international market for recycled products, and reducing greenhouse gas emissions.

As a result of the G7 Summit in Elmau in 2015, the Ise-Shima Summit in 2016 and the Toyama Framework for the Material Cycle, the Italian G7 Chairmanship in 2017 actively contributed to this process, facilitating the adoption of the Bologna work plan to develop common actions with the point of view of the efficiency of resource use and the circular economy [8].

The European Union estimates the potential reduction in production costs related to waste prevention and reuse of materials at 600 billion euros per year and estimates the positive effect of the transition to business models at about 700 thousand new employees [9]. Important figures that overcome the seeming dichotomy between sustainability and growth, which has always been a critical factor holding back the transformation of the production system towards more sustainable models. Therefore, it is no coincidence that this issue is being studied by such a supporter of innovation as Accenture [10], which came to the definition of a structure of values in the field of sustainable development, which determines the impact of the circular economy as the so-called "top line" (new products related to sustainability) and improving intangible value and "net profit" (reducing costs and improving risk management). According to experts, new opportunities for global business associated with the circular economy are estimated at about $ 4.5 trillion by 2030. These opportunities are directly related to the elimination of negative effects from the linear economy through the introduction of renewable energy sources and degradable materials, regeneration and provision of services for the optimization and sharing of resources, expanding the practice of recycling and recovery of energy [11].

The value of the circular economy, according to Accenture [12], is fully manifested in the elimination of the very concept of wasteful consumption and production. From this concept originate five "circular" business models that are functional for creating innovative products and services. "The circular supply chain" is the first of these models. It builds on innovation and research and offers access to renewable, recyclable or biodegradable resources instead of linear ones. In Italy, to give impetus to this phenomenon, the bioeconomy develops an annual turnover of about 330 billion euros (in Europe it is better only in Germany and France), and it employs 2 million workers [13].

The second model, more often associated with the concept of a circular economy, is the model of recycling and therefore reusing energy, materials and waste in production processes. Thinking about maximizing the value of an asset at the end of its life, it is necessary to rethink the design of new products in the name of modularity, definition of materials, durability: such giants as Nike and Ikea [14] are adopting this model.

Product lifecycle extension is the third possible business model: it starts with product restoration (additional capabilities—repair and updating of its functions and functionality), sale to the target audience. Google is applying this concept to its datacenters by placing surplus spare parts identified through an obsolescence management project in the aftermarket [15]. To date, about two million units have been sold in this way.

Sharing platforms that offer a "virtual" meeting place for product owners and individuals or organizations interested in using them are another example of how business in a circular economy one can do while reducing your environmental impact. This is the principle adopted on the Floow2 marketplace, a virtual B2B marketplace that links excess supply and demand for equipment and machinery, assets, personnel [16]. This means that each company can provide various organizations with free and non-profitable assets, such as equipment, production waste, buildings, employees, and so on. In 2018, more than 17 thousand users used the service of this company.

Finally, there is a model that has already become the paradigm of "product as a service" [17], according to which the manufacturer retains ownership of the asset and constantly sells its services, providing the services necessary for its use throughout its entire life cycle, including regeneration and processing. Thus, according to experts, the impact of the product on the environment will be reduced by (20-50)%. Also in this case, there is no shortage of examples: Signify (formerly Philips Lighting) sells lighting hours, not lamps, Michelin [18] sells kilometers traveled with its tires, SafeChem offers transport, storage and handling of chemical solvents as an alternative to buying the same.

Currently, in the world, including Italy, many projects are successfully functioning and aimed at developing a circular economy and smart waste recycling. One of the examples of such innovative projects is a citrus waste recycling company. We produce 700,000 tons of citrus waste every year. What if we don't throw them away? This question is being asked by two Sicilian entrepreneurs who have developed a patent that turns orange waste products into silk. How is this possible? The fabric is made from citrus pulp, that is, wet residues that remain at the end of the industrial production of citrus juice and which can no longer be used, but only thrown away as waste. Thanks to nanotechnology applied in creation of vitamin-rich clothing, the two entrepreneurs have been able to harness the potential of cellulose to extract citrus pulp suitable for spinning, thus turning industrial waste into the highest quality material for fashion. Fiber is the only brand in the world to produce the world's first eco-friendly citrus fabric [19].

Another interesting example of a business aimed at developing a sustainable economy is the AQUAFIL company [20].

According to the UN, 600,000 tons of fishing nets are dumped annually [21], which is 10% of the plastic waste found in the oceans.

Circular economy principles are perfectly matched to AQUAFIL, the world champion of Econyl, a nylon made not from petroleum (which produces carbon dioxide and waste), but from recycled materials that have reached the end of their life, including fishing nets from the sea, and rugs and carpets made in North America.

The company produces polyamide yarn, which is mainly used (over 70%) for the manufacture of carpets and floor coverings, including for cars. The rest of the production is nylon, from which the textile industry produces technical and sportswear.

Aquafil has invested 140 million euros over the past ten years to develop nylon recycling technology; a chemical recycling technology based on depolymerization, which makes it possible to make a new fiber completely identical to the original.

The biggest problem for the company remains the so-called "reverse logistics": trash search and organization of transport. On the other hand, the philosophy of the recycling company is not to wait for the waste to arrive at you, but to look for it.

Today, with 3,000 employees and 16 factories on three continents, Aquafil is the ninth nylon manufacturer in the world.

Separate waste collection, especially in large historical cities of Europe, where it is necessary to create a separate waste collection system from scratch and integrate it into the historical context, is a pressing issue. Separate waste collection in historic centers has its own characteristics that affect the lives and habits of townspeople and affect the very image of the city. The new outdoor devices are pieces of outdoor furniture that must guarantee the same performance standards as door-to-door segregated waste collection, without affecting the environment.

Thus, was born Citybin [22], a new system for the separate collection of plastic, organic, glass, aluminum and tin packaging, which combines a classic door with a door that remains active for paper and solids. Designed by Giulio Iacchetti, it combines a number of aesthetic and functional qualities.

Citybin is designed for convenient use by all citizens. It prevents overfilling, informs operators when saturation is reached, and optimizes emptying operations by minimizing the use of collection vehicles. The colors are designed to blend in with the buildings of the historic center, becoming an element of outdoor furniture. It does not have days and deadlines for delivery, with the help of a simple card, every citizen will be able to hand over waste when he considers it most convenient. If segregated collection is difficult enough, it is as difficult as what happens later, when waste is collected, loaded onto a vehicle and then disappears from view. The waste is transported to various sorting plants, where it is further separated, and from there, the waste is sent to centers where it is sent for regeneration. It is at this stage, hidden from citizens, that the sustainability of the process can potentially emerge: in fact, travel can carry waste over long distances—even abroad—thus having an impact on the environment.

However, it is clear that while recycling is a good practice, it also has economic and environmental costs. And this is not always an efficient process: if, for example, glass and aluminum can be recovered 100%, today only slightly more than 40% of the collected plastic is recycled.

A circular economy involves actions throughout the lifecycle of materials to complete cycles and improve resource efficiency. It is no longer just a "recyclable economy", but the emphasis is on the entire value chain, which includes many public and private actors and fosters effective collaboration processes and new business models. Industrial symbiosis (or industrial metabolism) includes traditionally separate industries with an integrated approach aimed at promoting competitive advantages through the exchange of materials, energy, water and/or by-products [23]. Symbiosis acts as a systemic eco-innovation tool for efficient use of resources and engages heterogeneous companies through the creation of resource sharing networks, including through appropriate supply and demand matching platforms.

3 Conclusion

The importance of a circular economy is reflected in the change in the concept of wasteful consumption and production, towards its weakening. From this concept originate five "circular" business models that are functional for creating innovative products and services. These models include "circular supply chain", circular economy concept, product lifecycle extension, platforms for sharing, "product as a service".

The circular economy is one of the main development models for the coming years. Some organizations have been studying this issue for a long time, create specific scenarios and projects for the next decades: for example, the 2030 Agenda for Sustainable Development, as well as the Circular Economy Package, the proposal launched by the EU and European Commission for Decarbonization by 2050 [24].

The circular economy represents an "epoch-making" change in perspective and, at the same time, the systemic ability to combine the needs of sustainable growth and combating waste with the needs of competitiveness and innovation. Moreover, it is worth noting that companies that use green and innovative technologies in their production process, as well as competently using production waste, they can not only increase profits, find a new niche in the world market, but also contribute to the development of a sustainable green economy. Thus, modern production facilities not only create new jobs, but also develop the territory where they are located, while reducing the environmental burden. Green technologies also allow for the greening of existing industries, improving the socio-economic status of industrial cities—all this leads to sustainable economic development in terms of ecology.

References

1. Meadows D, Randers J, Meadows D (2007) Limits to growth: the 30-year update, transl from English. ICC "Akademkniga", 342 pp
2. Waste. Second life. Electronic resource. Access mode: https://plus-one.rbc.ru/othody/. Accessed 26 June 2020
3. UN. "Electronic resource". Access mode: https://www.un.org/development/desa/en/news/population/world-population-prospects-2019.html
4. European Commission Making sustainable consumption and production a reality Luxembourg: Publications Office of the European Union 2010. 32 pp. ISBN 978-92-79-14357-1. doi:10.2779/91521
5. OECD (2016) Policy Guidance on Resource Efficiency, OECD Publishing, Paris. "Electronic resource". Access mode: https://doi.org/10.1787/9789264257344-en. Accessed 14 July 2020
6. Wilts CH. Germany on the Road to a Circular Economy. Friedrich-Ebert-Stiftung Publisher: Division for Economic and Social Policy Godesberger Allee 149/ D-53175 Bonn
7. G8 Environment Ministers Meeting 2008 Kobe 3R Action Plan. Electronic resource. Access mode: https://www.env.go.jp/en/focus/attach/080610-a5.pdf. Accessed 14 July 2020
8. G7 Bologna Environment Ministers' Meeting Bologna, Italy 11–12 June 2017 Communiqué. Electronic resource. Access mode: http://www.g7italy.it/sites/default/files/documents/Communiqu%C3%A9%20G7%20Environment%20-%20Bologna/index.pdf. Accessed 07 Oct 2020
9. Circular Economy Closing the loop AN ambitious eu circular economy package. "Electronic resource". Access mode: https://ec.europa.eu/commission/sites/beta-political/files/circular-economy-factsheet-general_en.pdf. Accessed 10 July 2020
10. Circular Advantage Innovative Business Models and Technologies to Create Value in a World without Limits to Growth 2014 Accenture. Electronic resource. Access mode: https://www.accenture.com/t20150523T053139__w__/us-en/_acnmedia/Accenture/Conversion-Assets/DotCom/Documents/Global/PDF/Strategy_6/Accenture-Circular-Advantage-Innovative-Business-ModelsTechnologies-Value-Growth.pdf. Accessed 14 July 2020
11. Future possibilities report 2020 2020 Centennial Lab, Ministry of Cabinet Affairs and the Future, UAE P.O. Box 21200, Dubai, United Arab Emirates ISBN 978-1-64970-314-9
12. Accenture (2019) Mining new value from the circular economy. Electronic resource. Access mode: https://www.accenture.com/il-en/iwnsights/naturalresources/mining-new-value-circular-economy. Accessed 14 July 2020
13. https://ec.europa.eu/research/bioeconomy/pdf/bit_en_strategy_2019.pdf. Accessed 14 July 2020
14. IKEA Electronic resource. Access mode: IKEA reports 88% recycling and recovery rate https://www.letsrecycle.com/news/latest-news/ikea-reports-88-recycling-and-recovery-rate/. Accessed 16 July 2020
15. Environment Projects Once is Never Enough March 2018. Electronic resource. Access mode: https://sustainability.google/progress/projects/circular-economy/. Accessed 16 July 2020
16. FLOOW2, World's Reset Button: The first B2B sharing platform Sustainable and social benefits through asset sharing. "Electronic resource". Access mode: https://ec.europa.eu/environment/ecoap/scalingcircularbusiness/solution/floow2_en. Accessed 8 May 2020
17. Moving from Product to Product-as-a-Service "Electronic resource". Access mode: https://www.engineering.com/IOT/ArticleID/18205/categoryId/44/Moving-from-Product-to-Product-as-a-Service.aspx. Accessed 1 Aug 2020

18. Chloé R, Frédéric D, Ulaga W (2010) Michelin Fleet Solutions: From selling tires to selling kilometers. HEC Paris case no. 510-103-1, 2010
19. Orange fiber sustainable fabrics from citrus juice by-products "Electronic resource". Access mode: http://orangefiber.it/en/. Accessed 5 Aug 2020
20. AQUAFIL company website. Electronic resource. Access mode: https://www.aquafil.com/who-we-are/profile/. Accessed 9 June 2020
21. Site Food and Agriculture Organization, FAO Electronic resource. Access mode: http://www.fao.org/in-action/globefish/fishery-information/resource-detail/en/c/388082/ Accessed 16 Nov 2020
22. Citybin Electronic resource. Access mode: https://www.mantovaambiente.it/citybin#intro-citybin. Accessed 6 Aug 2020
23. Chertow MR (2000) Industrial Symbiosis: Literature and Taxonomy," Energy and Environment Annual Review 2000
24. FINAL REPORT of the High-Level Panel of the European Decarbonisation Pathways Initiative, European Commission Directorate-General for Research and Innovation, European Commission B-1049 Brussels, 2018

Legal Support of Sustainable Development in Modern Societies

The Main Directions and Forms of International Cooperation in the Field of Illicit Drug Trafficking

Irina A. Uvarova

Abstract Current directions of international cooperation in the investigation of crimes related to illicit drug trafficking are considered in the paper. Speaking about international cooperation in the struggle against crime in general, it is noted that such activities are an important area of international relations that ensures domestic and international legal order, as well as national and international security. The paper considers narcotization as one of the main threats to the national security of society and an individual state. Implementation of directions and forms of international cooperation in the detection, solving and investigation of crimes in the field of illicit drug trafficking is aimed at solving a number of issues, namely: drug crime rate studies; organization of prevention of illicit drug trafficking; direct struggle against drug crimes; special rules on regulation of the turnover of narcotic drugs and psychotropic substances.

Keywords Drug crime · Drug addiction · Narcotic drugs · Illicit drug trafficking · Forms of international cooperation · Directions of international cooperation · Struggle against illicit drug trafficking

Despite the consolidation of law enforcement agencies and active combating transnational drug crime, narcotization in Russian society poses a serious threat to national security and is one of the main factors that undermine the country's demographic and socio-economic potential.

In the national security strategy of the Russian Federation (approved by Presidential Decree No. 683 of 31 December 2015) stipulates that the main threats to state and public security are … the activities of criminal organizations and groups, including transnational ones, related to illegal trafficking of narcotic drugs and psychotropic substances, weapons, ammunition, explosives, organization of illegal migration and human trafficking.

I. A. Uvarova (✉)
Moscow Witte University, 12, 2nd Kozhukhovsky Pas., Bldg. 1, Moscow 115432, Russia
e-mail: iuvarova@muiv.ru

A. V. Semenov and I. A. Sokolov (eds.), *Sustainable Development: Society, Ecology, Economy*, Earth and Environmental Sciences Library,
https://doi.org/10.1007/978-3-030-73110-6_12

The drug business is actively transforming, having a devastating impact on the social and economic areas of life of society and the state. The struggle against illicit drug trafficking is one of the priorities in the implementation of the country's development strategy.

An important component in building a modern healthcare system, as well as a healthy society in general, is the setting of special rules that regulate the circulation of narcotic drugs and psychotropic substances.

At present, there is not a single state that does not realize the danger and harmful consequences for mankind of illegal drug trafficking. Drug abuse and drug addiction of a fairly large part of the population can already be considered as a national disaster.

Not a single state can cope with the rapid spread of illicit drug trafficking on its own. This requires international cooperation and a collective solution to this problem.

Currently, joint measures have been developed and are being actively used to combat the illegal distribution of narcotic drugs. Within the framework of inter-state cooperation, the following legal mechanisms are applied [9]:

1. the Convention mechanism (or treaty and legal), which provides for the conclusion of special agreements regulating relations in this area [7];
2. an institutional mechanism that includes cooperation within international bodies and organizations.

It is worth paying attention to international legal acts in this area and analyzing the history when considering the directions of international cooperation in the field of illicit drug trafficking.

The first international Convention on narcotic drugs was adopted in 1912 in Hague and settled a number of principles that were later reflected in international agreements. The Geneva Convention of 1936 settled international criminal prosecution for crimes related to the distribution of narcotic drugs. In 1946, The Commission on Narcotic Drugs was established to organize and coordinate international activities in the field of illicit drug trafficking.

The following conventions are currently important:

Single Convention on Narcotic Drugs (1961)[1];

UN Convention against illicit traffic in narcotic drugs and psychotropic substances (1988);

UN Convention against transnational organized crime (2000)[2];

Agreement between the member States of the Shanghai cooperation organization on cooperation in combating illicit trafficking in narcotic drugs, psychotropic substances and their precursors (2004).

The fundamental anti-drug agreement in the fight against illicit traffic in narcotic drugs is the United Nations Convention against Illicit Traffic in Narcotic Drugs

[1]Single Convention on Narcotic Drugs of 1961, as amended in accordance with the 1972 Protocol on amendments to the Single Convention on narcotic drugs of 1961 (concluded in new York on 30 March 1961)// LRS "ConsultantPlus".

[2]Convention against Transnational Organized Crime "ConsultantPlus". (adopted in new York on 15 November 2000 by Resolution 55/25 at the 62nd plenary session of the 55th session of the UN General Assembly) (with ed. 15 November 2000)// LRS "ConsultantPlus".

and psychotropic substances,[3] the purpose of which "... is to promote co-operation among the Parties so that they may address more effectively the various aspects of illicit traffic in narcotic drugs and psychotropic substances having an international dimension" (part 1 of article 2 of the Convention).

Considering the forms of international cooperation in the field of illicit drug trafficking, it should be noted that States provide each other with mutual legal assistance in the detection, solving and investigation of crimes and judicial proceedings. Such assistance may be provided in the following cases [4, 6]:

- to collect evidence or testimony;
- to get acquainted with the materials of the trial;
- for executing searches and seizures;
- for inspection of objects and sites;
- to provide information and evidentiary items, etc.

Other forms of cooperation include:

1. Transfer of materials of criminal proceedings. In this case, the possibility of mutual transfer of materials for criminal prosecution of offenses is considered when it is established that such transfer is carried out in the interests of justice. However, the question arises: what is the difference between criminal prosecution in accordance with paragraph 55 of article 5 of the Criminal Procedure Code of the Russian Federation and criminal prosecution in the framework of international cooperation? Considering the norms of article 55 of the Criminal Procedure Code of the Russian Federation, it is not possible to distinguish them from the institute of legal assistance in criminal cases, since in such assistance, it is possible to discover the features of criminal prosecution, for example, when requesting the interrogation of a suspect or accused. Therefore, in our opinion, it is necessary to be of the view of those scientific proceduralists who understand "criminal prosecution" as "... part of international legal cooperation in the field of criminal justice, which has a special legal regulation, legal content, goals and objectives, since otherwise, in certain circumstances, mutual legal assistance, extradition and other areas of interstate interaction in criminal cases should also be considered criminal prosecution[4]". Klevtsov K. K.[5] notes that the transfer of criminal prosecution (legal proceedings) is aimed at providing the full scope of criminal proceedings by one state to another in order to bring the suspect or accused to criminal liability, or initiating such proceedings in the state where such a criminally prosecuted person is, if copies of the case materials are transferred, since it is not possible to ensure that this participant

---

[3]United Nations Convention against illicit traffic in narcotic drugs and psychotropic substances (concluded in Vienna on 20 December 1988)// LRS "ConsultantPlus".

[4]Bykova E. V., Reshetnikova T. A., Vyskub V. S. Transfer of Criminal Prosecution (Legal Proceedings): Legal Regulation And Some Problems. Library of the Criminalist. Scientific Journal. 2015. - No. 3. - P. 295.

[5]Klevtsov K. K. The Concept and Legal Nature of the Transfer of Criminal Prosecution (Proceedings). International Criminal Law and International Justice. - 2018. - No. 6. - P. 11–14.

in the criminal process is in the state where the original criminal case is being investigated.

2. Establishing and maintaining communication channels between agencies and services to enable rapid and reliable exchange of information that relates to all aspects of offences.
3. Cooperation in the investigation of offences that are of an international nature in order to:
   - identification, finding out the whereabouts and activities of persons suspected of participating in offences;
   - transfer of income or property, which were obtained as a result of such offenses;
   - movement of narcotic drugs, psychotropic substances of the United Nations Convention against Illicit Traffic in Narcotic Drugs and Psychotropic Substances, and means used or intended for use in the commiting of such offences.
4. Creating specialized groups that ensure the safety of individuals and their operations.
5. Providing, in certain cases, the necessary samples of substances for research and crime investigation.
6. Facilitate effective interaction between agencies and services for the exchange of staff and experts.

Thus, considering the modern anti-drug legislation based on international instruments, it is worth noting that currently greatly enhance the fight against drug addiction and drug abuse, has established a comprehensive legal system for the boundaries deterrence of the negative social phenomena.

References

1. Single Convention on Narcotic Drugs of 1961, as amended in accordance with the 1972 Protocol on amendments to the Single Convention on narcotic drugs of 1961 (Concluded in new York on 30.03.1961). LRS “ConsultantPlus”
2. Convention Against Transnational Organized Crime “ConsultantPlus”. (adopted in new York on 15.11.2000 by Resolution 55/25 at the 62nd plenary session of the 55th session of the UN General Assembly) (with ed. from 15.11.2000). LRS “ConsultantPlus”
3. United Nations Convention against illicit traffic in narcotic drugs and psychotropic substances (concluded in Vienna on 20.12.1988). SPS “ConsultantPlus”
4. Altunin TV (2017) Separate International anti-drug agreements of the Russian Federation as sources of rules for the turnover of narcotic drugs and psychotropic substances: criminal-legal aspect. The Modern Law 6:121–126
5. Bykova EV, Reshetnikova TA, Vyskub VS (2015) Transfer of Criminal Prosecution (Legal Proceedings): Legal Regulation and Some Problems. Criminalist’s Library Sci J 3:295
6. Gotchina LV (2014) International legislation in the field of prevention of drug crimes committed by minors, Youth and in Relation to Them. Lex Russica 5:594–606

7. Gusev AV (2010) Problems of implementation of international agreements of interdepartmental nature of law enforcement agencies of executive power. Russian Law J 5:102–110
8. Klevtsov KK (2018) The concept and legal nature of the transfer of criminal prosecution (Proceedings). Int Criminal Law Int Just 6:11–14
9. Lyudkevich YuN (2008) Experience and problems of international cooperation of law enforcement agencies of Russia in the field of prevention of illicit traffic in narcotic drugs and psychotropic substances. Soc Law 2. LRS "ConsultantPlus"

Liability Insurance for Causing Harm to Life and Health as a Legal Remedy for the Protection of Legitimate Interests in Conditions of Modern Society Development

Irina N. Romanova

Abstract Under certain circumstances, any type of activity may cause harm to life and health, and, therefore, give rise to an obligation to compensate it. This problem is particularly relevant in the context of scientific and technological progress, associated with the active operation of dangerous facilities, the implementation of new technologies. The purpose of the paper is to study the features of the liability insurance mechanism for certain categories of goods producers and executing agents (service providers), considered as a means of effective monitoring of their activities, ensuring the protection of the rights of services consumers, especially in those spheres of national economy where there are no clear and effective mechanisms of state control and supervision, which is of particular importance in modern conditions. The methodological base of the investigation was made up as of general methods of scientific cognition, such as dialectical method, induction, deduction, and special methods: formal-logical, comparative law, modeling. In the course of the research, we based on a study of scientific points of view, analysis of regulatory acts. It allowed us to conclude that liability insurance for causing harm to life and (or) health is mainly based on the compulsory insurance segment, which currently includes compulsory civil liability insurance of the owner of the dangerous facility for damage resulting from an accident at this facility, of the owner of vehicles and the carrier, each of which is quite controversially developed, requiring constant adjustment of the legal prescriptions.

Keywords Liability insurance · Compensation payments · Insurance event · Insurance cover · Interests of the insured

I. N. Romanova (✉)
Moscow Witte University, Branch in Ryazan, 5, Yablochkova Ave, Ryazan 390023, Russia
e-mail: iromanovarzn@bk.ru

A. V. Semenov and I. A. Sokolov (eds.), *Sustainable Development: Society, Ecology, Economy*, Earth and Environmental Sciences Library,
https://doi.org/10.1007/978-3-030-73110-6_13

1 Introduction

The specificity of liability insurance determines the high potential of the idea of using compulsory liability insurance mechanisms in the form of a substitute (alternative) for certain types of state control (supervision), since it provides the latter's timely and adequate verification and assessment by the insurer of compliance of the policyholder's activity with mandatory requirements due to disinterest in the occurrence of an insured event. Liability insurance is currently one of the quite common types of insurance, which, on the one hand, is caused by increasing interest of business entities in this method of risk management, and, on the other hand, the expansion of the scope of compulsory non-State insurance.

In Russia, there is a clear trend towards internal differentiation of civil liability insurance, which directly affects the insurance products offered by insurers. At the same time, there is a trend towards detalization of liability insurance of certain categories of entities, which leads to the emergence of specific rules for liability insurance of builders, architects, planners, forwarding agents, customs representatives, air carriers, shipowners to third parties, directors and other heads of executive bodies.

Liability insurance for causing harm to life and (or) health affects the implementation of tort liability functions, reducing the power of prevention provided by liability risk and creating a dilemma regarding the priority of punishing the guilty person or protecting the property interests of the victim. At the same time, the emphasis is shifted from the fault of the causer of harm to determining the amount of compensation, and insurance itself acts as an element of positive responsibility, stipulating participation in the insurance system regardless of the actual harm.

2 Models of Insurance Cover Against the Risk of Liability for Causing Harm to Life and (or) Health

Based on the analysis of foreign legislation, specialists identify several models for the implementation of insurance cover for liability insurance for causing harm. The first of them—the deterrence model—is based on the distinction between liability for compensation of harm and insurance obligation, each of which is performed independently in a certain sequence. A similar practice has spread in some countries of Western Europe. In particular, in France, the obligations of the insurer under a liability insurance contract arise only if the fact of payment of compensation to the victim is confirmed, the amount of which consequently determines the amount of insurance payment, although payment can also be made directly to the victim in the amount of uncovered damage. Since in this case the harm causer settles with the victim independently, without any relation to his insurance contract, as the insured, he simultaneously acts as the beneficiary to whom the insurance payment provides compensation for the damage caused by the insured event, as measured by the payment made to the victim. The theoretical basis for building such relations is formed by the economic doctrine of the deterrent function of tort law, on the basis of which tort responsibility should have an active preventive and educational effect on the personality of the offender, and, therefore, the offender should be brought to justice regardless of whether the corresponding risk is provided with insurance. Moreover, the presence of the latter is more likely to be considered as an additional problem, since it weakens the effect of prevention, stipulated by the existence of responsibility.

The compensation model, focused on the maximum protection of victims, on the contrary, involves the simultaneous implementation of a mechanism for bringing the tortfeasor to responsibility and the performance by the insurer of the duty to pay insurance compensation to the victim who is the beneficiary in this case. In Russia, as in the most of countries, this particular model obtained legislative consolidation, however, in some countries, in particular, in New Zealand, the legislator went even further by creating a model of compensation without fault in order to ensure equal care for all victims through the introduction of special taxes for persons engaged in activities that are potentially dangerous to others, or imposing on them the obligation to insure potential damage. At the same time, as a rule, compensation for harm to victims is carried out regardless of the fault of both the causer of the harm and the victim in a predetermined minimum amount.

It should be noted that the question of the need to develop liability insurance for causing harm in Russia was repeatedly raised in program documents that determine the main directions of insurance development. Already in 1998, the task of developing draft laws on compulsory liability insurance of vehicle owners, employers, as well as certain categories of manufacturers (operators, service providers) was set. This idea was developed in the Concept for the Development of Insurance in the Russian Federation, where along with the introduction of types compulsory insurance of facilities subject to significant risks and losses, citizens and legal entities that suffered significant damage caused by natural disasters, accidents and catastrophes (liability insurance of vehicle owners, insurance of industrial facilities against man-made accidents, property insurance against fires and natural disasters, insurance against damage as a result of major accidents during the movement of dangerous cargo), the relationship between the development of entrepreneurship and liability insurance in the production of goods, the performance of works, the provision of services, as a result of which harm may be caused to third parties, was indicated.

The object of insurance in such cases is the property interests of the policyholder (insured person) associated with the risk of liability for harm to life, health or property of citizens, legal entities, public law entities (beneficiaries) due to defects of the goods, services provided, works performed, manufactured and (or) sold by the policyholder (insured person), as well as providing them with inaccurate or insufficient information about the goods, work, service. Sometimes, property interests associated with the risk of unforeseen legal and other costs caused by the claims (lawsuits, complaints) for compensation for harm are also recognized as such. The risk of liability for such harm is insured under an insurance contract, liability for causing harm to the environment [2].

When determining the object of insurance, insurers focus on such characteristics of the goods (works, services) as: mass or serial production (responsibility for the quality of a single and unique product can be accepted for insurance only by agreement of the parties); unambiguous requirements for product quality indicators that meet international standards or current legislation; the policyholder's right to manufacture and sell goods (works, services); implementation of documented measures to ensure the quality management system of the goods.

An insured event, as a general rule, is the occurrence of the obligation of the policyholder (insured person) to compensate for harm caused to the life, health or property of consumers and/or other persons, as a result of unintentional provision of false or insufficient information about the product (goods, work, service), as well as the presence of hidden constructive, prescription or other unintended defects in the product (goods, work, service), including due to unforeseen defects or special properties of the materials used, tools, hardware failure, etc. Moreover, the insured risks differ significantly from those inherent in professional liability insurance [3]. First of all, the scope of their occurrence is distinguished, which in the case of professional responsibility is limited to the performance of work and the provision of services. The moment of harm is also important in the case under consideration, as it may occur only after delivery of the goods (work, service), while the risk of professional liability accompanies the entire period of activity. The nature of the relationship between the parties also differs in principle. In cases where there is a risk of causing harm with low-quality products, the presence of direct contact or contractual relationship between the tortfeasor and the victim is not a necessary condition for damage, while professional risk affects the customer who directly enters into a contractual relationship with the contractor [7].

Insurers pay special attention to circumstances that are not covered by insurance cover, among which one can highlight: a group of force majeure circumstances (military operations, nuclear explosion, etc.); deliberate consumer actions, including the use of goods and work results for experimental or research work; the deliberate actions of the policyholder (insured person) related to the violation of the established procedure for the production and sale of goods, works and services; events recognized as insured events for other types of insurance (for example, manufacturer insolvency); other circumstances determined by the insurer as not covered by insurance, for example, the impossibility of withdrawing, recalling, checking, repairing, replacement, equipment, eliminating or removing goods, works, services of the policyholder (insured person), affiliation of the policyholder and beneficiary, etc.

The insurance amount is established by agreement of the parties and is the limit of insurance compensation for all insurance cases occurring during the period of validity of the insurance contract. It may, however, contain limit liability per single victim, per single insurance event, or for a separate category of goods (works, services).

3 Dynamics of Insurance Cover Against the Risk of Liability for Causing Harm to Life and (or) Health

One of the most important issues is the period of insurance coverage. On the basis of the principles applied in foreign insurance markets, the date of the insured event is the presentation of a claim to the policyholder (insured person) or causing harm to the consumer during the term of the insurance contract. Therefore, AIG Russia Insurance Company associates the provision of insurance cover with the presentation to the policyholder of a court decision or an officially declared property claim during the period of validity of the insurance contract, unless the latter provides for a specific period of time during which the insurer accepts the claims arising due to the infliction of harm for consideration and settlement, to the extent that the event which caused it occurred within the contract term.

A slightly different approach is implemented by Russian insurers, who usually consider the fact of causing harm during the term of the insurance contract (and if this is stipulated by the contract, they would do it during the period determined by the parties after the contract has expired) and presenting claim for compensation for harm during the term of the contract, if it did not establish that insurance applies to claims made, inter alia, in a certain period after the termination of the contract as the totality of the circumstances giving the right to obtain insurance compensation. The situation is aggravated by the existence of reservations that during the term of the insurance contract the products that caused harm should be manufactured (sold), although at the same time, by agreement of the parties, the contract can applicable to products that were manufactured and/or sold (fulfilled, provided), in a certain period before the commencement of the contract. However, Rosgosstrakh is reasonably limited by the terms of the sale of the goods, considering that it can take a lot of time from production to the moment the product is sold to the consumer. However, in general, the implemented approach to determining the insurance period is objectionable, since it reduces the level of insurance cover of both the policyholder and the beneficiary.

Speaking of liability insurance for causing harm, it is worth paying attention to liability insurance of the employer, whose property interests in such cases are associated with the risk of his liability for causing harm to life, health and/or property of citizens, in the performance of their labor duties or due to the performance of works (services) under a civil contract concluded with the policyholder. It should be noted that in the absence of regulatory requirements, insurers determine the moment of occurrence of the insured event differently [6].

In this case, again, we are talking about unintentional errors, negligence and omissions of the policyholder or a person whose liability risk is insured, which in this case is understood as an unintentional violation of regulatory documents and requirements established by current legislation to ensure safe working conditions and labor protection, conduction of specific types of work and services or a violation of the safety of goods, works, services, which were manufactured/operated/serviced by policyholder, including fire and security alarm systems.

Liability insurance for causing harm is mainly developed through the compulsory insurance segment. In particular, in accordance with Strategy 2020, a number of measures were taken to develop compulsory civil liability insurance for the owner of a dangerous facility in terms of setting insurance rates, maintaining statistics of insurance events for calculating insurance premiums using a special information system, expanding the list of potentially dangerous facilities subject to insurance (which were supplemented by passenger conveyors), insured events (in particular, leakage, emission of hazardous substances, collapse of the rock (masses of the rock) were also recognized as an accident; operations related to technical re-equipment, overhaul), the amount of liability of the insurer (which was doubled) and the procedure for determining the amount of damage.

Not without the reason it is believed that further expansion of the list of compulsory types of insurance may result in serious negative consequences in the future [8], in particular, increased tension among insurers caused by dissatisfaction with tariffs and insurance rules, increased desire for fraud on the part of insurers clients and an increase in the loss ratio of insurance operations, which was especially manifested in the field of motor insurance. As a result, the development of liability insurance by expanding the list of types of "imputed" insurance requires increased attention to protecting the rights of policyholders as a special category of business entities [4], although, according to experts, the expansion of this market segment is largely due to their low insurance culture, not allowing them to take full advantage of the practice of voluntary liability insurance existing abroad, which is, among other things, stimulated by large expenses on settlement of damages which are not covered by the active participation of the state [5].

It should be noted that for the emergence and development of insurance relations, not only the characteristics of the vehicle or dangerous facility used are important (in particular, the obligation to conclude an insurance contract does not arise for metro transport operations), but also the nature of the activities carried out. Thus, the courts of several instances, evaluating the evidence presented in their totality and mutual relationship (including evidence of the initial transfer of a Japanese citizen from one side to another), came to the conclusion that there was no obligation on society to conclude a carrier civil liability insurance contract for causing harm to life, health, property of passengers due to the fact that a foreign citizen was a specialist who performed work on a fishing vessel, which indicates the absence of mandatory features of transportation of passengers by sea. It is noteworthy that in relation to educational institutions that transport children with their own buses, enterprises that transport employees with their own buses, and other categories of non-commercial carriers, this issue was resolved positively and currently (since January, 2018), they are primarily responsible for the damage caused to passengers during transportation [1].

4 Conclusion

In the liability insurance system for causing harm, two groups of legal relations are clearly distinguished, differing in subject composition, the procedure for determining the list of insurance cases, insurance rates and the formation of a guarantee system for recipients of insurance payments, depending on whether it is compulsory (civil liability insurance of the dangerous facility owner for causing damage as a result of an accident at a dangerous facility, of a vehicle owner, and of a carrier) or voluntary.

Liability insurance for causing harm to life, health or property is carried out by business entities on their own initiative or by direct prescription of the law, which is mainly associated with the implementation of activities with a potential risk of harm (operation of dangerous industrial facilities and other sources of increased danger). In the latter case, the conclusion of an agreement is a prerequisite for the legality of the implementation of the activities provided for by law (provision of services for the transportation of passengers, operation of dangerous facilities).

Liability insurance for causing harm is mainly developed through the compulsory insurance segment, which currently includes compulsory civil liability insurance of the owner of a dangerous facility for damage resulting from an accident at this facility, of the owner of vehicles and of the carrier, each of which is quite controversially developed, requiring constant adjustment of regulatory requirements.

References

1. Batrova TA (2016) Problems of passenger insurance: law, logic, moral. The future of Russian insurance: estimates, problems, growth point. Proceedings of the XVII Int. scientific-practical conferences. Rostov-on-Don: Publishing house of SFedU, pp 290–295
2. Danilochkina MA, Savinsky RK (2008) Insurance of financial risks. Juridical and legal work in insurance, no 2, p 24
3. Matveychuk YuKh (2014) Entrepreneurial activity in Russia: risks and responsibility. Bus Manag Law 2:55–57
4. Makhdieva YuM (2014) Liability insurance in Russia: current status and development directions. Finan Credit 46(622):35–44
5. Moshkin AS, Polevoy EV (2016) Compulsory civil liability insurance of the owner of a dangerous facility: problems and prospects. Scientific and educational problems of civil protection, no 2, pp 120–123
6. Romanova IN (2019) Insurance in the field of entrepreneurial activity as a reducing factor of losses. Legal Concept 18(2):84–89
7. Sinishev DV (2004) Producer liability insurance for the quality of products, works, services (on the example of the consumer market): Author. dis. … cand. econ. sciences. SPb., 23 pp
8. Chemerisova EYu (2015) Entrepreneurial risk insurance contract: Author. diss. … cand. legal sciences. Kazan, p 9

Juvenalization of the State Policy of the Russian Federation in the Interests of Minors

Vasily S. Yurchuk

Abstract The policy of introduction of juvenile law and juvenile justice into the legal system of the state pursued by the Russian state is considered in the paper. This implies the formation of the state policy of the Russian state in this area, as an independent activity of state bodies and society. A characteristic of the concept of "juvenileization of law" is given. It is noted that at the present stage of the Russian state development, juvenile law is increasingly used, which more effectively protects the rights and interests of children with the help of legal resources. The main approaches that have developed in the Russian legal science in the field of research of legal phenomena, including in the field of juvenile law, are shown. The problems that exist in juvenile law of both objective and subjective nature are noted. One of these problems is the lack of systematization of juvenile law; it does not fit into the sectoral typology of law that has developed in Russia. The factors contributing to the juvenileization of law are highlighted, the author's position on the issue of his attitude to the concepts of "youth studies", "juvenile law", "juvenile justice" is noted. A characteristic of the concept of "the needs of a minor" is given. It is noted that this concept is fixed in certain principles and norms of Russian law.

Keywords Juvenile law · Children's interests · Juvenileization · Juvenile justice · The subject of juvenile law · The needs of the child · Youth studies · The children national strategy · The convention on the rights of the child

1 Introduction

The relevance of the study of the policy pursued by the Russian state in the field of protection of the rights and legitimate interests of children is due to the issues that the state solves to protect the life and health of the younger generation. This implies the formation of the state policy of the Russian state in this area, as an independent activity of state bodies and society. The negative trend that is taking place in the

V. S. Yurchuk (✉)
Moscow Witte University, 12, 2nd Kozhukhovsky Pas., Bldg. 1, Moscow 115432, Russia
e-mail: v.jurchuk@muiv.ru

A. V. Semenov and I. A. Sokolov (eds.), *Sustainable Development: Society, Ecology, Economy*, Earth and Environmental Sciences Library,
https://doi.org/10.1007/978-3-030-73110-6_14

sphere of protecting children's rights and ensuring their well-being is a feature of any state, and it is also evident in Russia. Among the factors inherent in modern Russian society, the following can be distinguished:

- the number of orphans detected annually is increasing;
- there is a growing number of cases of child abuse;
- juvenile delinquency increases;
- multiplies the number of single-parent families;
- the dynamics of development of such social phenomena as drug addiction and alcoholism among minors remains unfavorable.

The reasons for this situation were the shortcomings of the state policy on children, the lack of a state strategy in this matter, insufficient funding and support for children. An important factor was the weakening of the role of the family as a guarantor of economic security and development of children.

A new impetus to the state policy in the field of protection of the rights and interests of children was to be given by the "the 2012–2017 National Children's Interests Action Strategy", which assumed the juvenalization of law, which led to the further development of such concepts as" juvenile law", "juvenile justice".

The policy of the Russian state in the interests of minors is recorded in many legal documents. This, as noted above, is also the National Children's Interests Action Strategy 2012–2017, approved on 1 June 2012, where it was noted that children have the right to special care and assistance. This is also the Constitution of the country, which guarantees state support for the family, motherhood and childhood. This is the Convention on the Rights of the Child and other international instruments in the field of children's rights, where Russia has assumed a number of obligations to create an environment that is comfortable and friendly for children " [9].

The provisions of the Strategy indicated specific goals and directions of the Russian Federation's policy in the interests of children and created conditions for its legal support. The process of creating an environment comfortable and safe for a child's life is quite complex and contradictory itself. With the help of law, it can be quite successful, if the concept of "juvenile law" is taken into account. From the philosophical and legal point of view, the juvenalization of law is the evolution of a legal subject, its legal status as a minor.

History shows that the development of the social and legal status of children has a long period and it begins with their complete rightlessness, when the power of parents was unlimited. Gradually, the power of parents over children is limited, it turns into parental rights, which are controlled by law, and the responsibility of parents for its violation is established. Today, a minor is recognized as a special privileged subject of law, whose opinion in various situations must be taken into account, and he himself, his rights must necessarily be protected by law [12].

The historical contradictory development of juvenile law, the formation of its legal status has led to changes in the procedures and mechanisms for the protection of the rights of minors.

Solving legal problems related to the protection of minors, special attention is now paid to juvenile justice, which is directly related to minor children.

The introduction of juvenile law and juvenile justice as a more effective element of protecting the rights and interests of children into national legislation is an important issue in the legal system of the Russian Federation. Modern Russian legislation is increasingly using juvenile law, which protects the rights and interests of children more effectively with the help of legal resources.

The juvenalization of law, which is present in all processes occurring in legal ontology (legal consciousness, legal relations, legal culture), has a significant impact on the implementation of law, legal education and law studies. Discussions on the use of juvenile law and juvenile justice in the national legal system are important and demand great attention from the side of the legislator and enforcement officer. Presently, there are several legal concepts on the basis of which approaches in the field of scientific research of legal phenomena are built. Modern researchers of law, as a rule, consider it in the course of two approaches. The first approach is based on the classical model of legal positivism, which assumes a strict normativity of law, state monopoly in legislative, law enforcement practice, priority is given to legal enforcement. The second approach is based on the concept of natural law, which puts the natural rights and freedoms of the individual above the positive law created by the state.

Exploring the legal basis of juvenile law, the author believes that the Constitution of the Russian Federation, the Universal Declaration of Human Rights are the basis for juvenile law. Thus, the first article of the Universal Declaration of Human Rights States that all people are equal [2], while in the second article of the Russian Constitution it is stipulated: "the rights and freedoms are the Supreme value" [5] and article 17, second paragraph states that the fundamental rights and freedoms are inalienable and belong to everyone by birth [6]. All these provisions also apply to minors and form the basis of juvenile law being established in Russia.

It should be noted that the Universal Declaration of Human Rights and the Constitution of the Russian Federation have created the basis of the natural law concept, which must be taken into account during the juvenalization of law in the Russian Federation. However, the natural law concept itself expounded in the Basic Law is perceived differently by supporters of Russian positivist concepts, as well as by proponents of other concepts of legal understanding.

Representatives of this trend believe that the natural law provisions of the Russian Constitution are only the imperative of our time, they are not a legal norm, it is a tribute to the political situation. Some legal scholars believe that the natural law provisions written in the text of the Constitution of the Russian Federation do not reflect a specific legal reality, but only the natural law approach, which today dominates the national legal system.

Considering modern national law, law enforcement practice, we can note that there is a humanization of legal relations, and first of all, those that relate to the interests and rights of children. In Russia, the concepts of juvenile law and juvenile justice are firmly established. These concepts are introduced into the legal consciousness, legal ideology, legal relations in all branches of Russian law in law enforcement practice.

It can be stated that the juvenalization of Russian legislation in the interests of children is sustainable, however, its implementation in practice is far from perfect. There are still many problems of both objective and subjective nature.

In modern juvenile law, there are many different problems that are the result of a conflict of interests that exist in the social, economic, political spheres—between minors and other structures of society (the state, the adult population, parents, etc.).

The existence of law is realized in the legal relations of subjects of law. All of them have certain rights and legal obligations.

Here, the law, by means of normative acts, establishes the limits of these subjective rights, which end for a particular person where the rights and freedoms of another person begin.

Any processes taking place in Russian society affect the interests of minors, most of these relations require legal regulation. This, in turn, leads to an increase in the body of legislation regulating relations involving minors as the most vulnerable social group. The need to systematize these norms is an important issue for the state [7].

Juvenalization of law leading to the empowerment of minors, the emergence of various types of legal capacity, as a rule, leads to a decrease in the "paternalistic" rights of other subjects of law in relation to children as an object of legal influence, law enforcement.

Part of society is not ready for such legal relations, is not ready to give up the rights and obligations in favor of minors. In the system of law, as we see, there is a contradiction between parents and children, which is both objective and subjective. The juvenalization of law is an attempt to resolve this legal conflict, to reduce its severity, to show the stages of its resolution. The processes occurring in this area affect the understanding of the concept of "juvenile law", "juvenile legislation", "juvenile justice".

In Russia, there are different approaches and positions in the field of consideration of problems of juvenile law, juvenile legislation, juvenile justice, the theory of juvenile law. Thus, some legal scholars believe that in modern national legislation with its established branch of legal system, there is no need to create another branch of law as one of the eliments of legal reality. For example, V. S. Tolstoy believes that juvenile law is only a scientific discipline at the present. This statement follows from his thesis that the legal rules that regulate relations involving minor children are scattered across various branches of law. These rules are present in the Civil, Labor, Family, Criminal, Penal codes. The legislative technique has not allowed us to combine the norms concerning the legal status of minors into one legal document yet. Therefore, there is no reason to talk about the creation of a branch of juvenile law. Juvenile law in the legal system of Russia is a branch of science, but not a branch of law. It studies a large number of legal norms that are present in various branches of law and regulate the legal status of minors [14].

In Russian legislation, juvenile law is not framed as a branch of law, it still exists in an unsystematic form. Therefore, the question arises of what is the subject of juvenile law as a branch of science. In his theory Tolstoy V. S. believes that juvenile law is a synthesized, intersectoral legal theory, which is an important part of the theory of human rights. It can be noted that juvenile law, as a theory, is based on the practice that is represented by juvenile justice and this juvenile law has long been applied in other countries. Proponents of another approach believe that the existing juvenile law and juvenile justice are synonymous, therefore do not distinguish between these terms. Thus, A. V. Zaryaev, V. D. Malkov write, “juvenile law is a complex branch of the domestic legal system that combines legal norms regulating public legal relations, to which, at least, a child is a party (juvenile legal relations); forming the legal status of a minor as a participant in these legal relations; establishing mechanisms of legal protection of minors, as well as defining the powers and principles of the system of bodies and institutions aimed at the implementation and protection of the rights and legitimate interests of minors (juvenile justice bodies)” [3].

In our understanding, juvenile law is a natural, integral part of the law as such, the most important component of both the Romano-German and Anglo-Saxon legal system. Juvenile law is actively used by the existing legislation of any democratic legal state at the present.

Of course, the phenomena of juvenile law do not fit into the existing sectoral typology of law in our country.

We note that the sectoral typology of law that has developed in Russia does not leave room for juvenile law as a complex branch of domestic law.

The investigation of juvenile law as a synthesized branch of law, as it is conducted by A. V. Zaryaev, V. D. Malkov, E. A. Suntsova, I. V. Predeina, Yu. V. Nikolaeva and other legal scholars, can somehow solve the problems of its modern typology. However, this approach should take into account the positions as representatives of the domestic model of legal positivism and followers of natural law theory, which notes that the rights of minor children are natural and belong to them from birth. Therefore it should be concluded that juvenile law is also natural. However, juvenile law is implemented through juvenile legislation, and it is positive, created by the will of the legislator.

If the juvenile law, considered in the framework of natural law theory, acts as a unity of right and true, then the juvenile legislation recorded juvenile practice, which reflects the legal reality of a particular state.

Thus, the juvenile legislation is expressed in positive norms of law, where the ratio of due and existing, inherent in the natural-legal paradigm at a certain stage of development of the legal system of the state, is realized. The legal system of modern Russia is developing in line with this process, it is characterized by contradictions, it is influenced by political, ideological, and other factors.

As E. V. Kamenskaya notes, “in the legislative activity of the Russian Federation, there have been positive changes in terms of enhancing the interaction between the government and society, transparency of law-making decisions, the beginning of the regulatory legal framework formation”. This also fully applies to the juvenile law in Russia [4].

Modern official legal terminology studies has not yet formed an idea of the system of categories, concepts, terms that can be used in more effective investigation and protection of the rights and interests of children.

In many countries related to the Romano-Germanic and Anglo-Saxon families of law, a specific group of terms is used to designate the life of the youth society. The core of this group is formed by the philosophical, social and humanitarian category of “juvenileology”, which designates everything that is associated with childhood, adolescence, and with minors [13].

Most scholars view juvenology as a science of youth, although there is another approach according to which juvenology is the science of the rights of a child of a specific age, when partial or full legal capacity occurs, and this is in the time interval from 6 to 18 years. The term “juwenalia” in the process of its development is introduced into scientific parlance by a number of other Sciences. Such concepts as” juvenile sociology”,” juvenile anthropology”,” juvenile psychology”, etc. are firmly included in the lexicon of modern science.

Jurisprudence also uses terms such as “juvenile law”, “juvenile legislation”, “juvenile justice”, “juvenile justice procedure” very widely.

The term “Juvenile”, meaning youth, minority, in the history of Russian law was not used so often. Thus, for example, in the XIX—early XX centuries in scientific and professional vocabulary, it was almost never used. During this period, the concepts of “special courts for youth”, “juvenile courts”, “justice in relation to minors” and a number of similar, more understandable to the public terms were in use [8].

In the Soviet period, the term “juvenile justice”, “juvenile law”, “juvenile legislation” was not used, although in the first years of the Soviet Union in the country there were courts for such cases. And, only in modern Russia, since the 1990s, the terms with the concept of “juvenile” have been actively introduced. This issue was studied particularly intensive in the end of XX – beginning of XXI century. Hundreds of PhDs were defended, a significant number of various conferences were held, monographs, textbooks, training manuals were published.

Gradually, interest in the problems of legal juvenology began to decrease. In our country, over the past decade, in the public consciousness there has been a rejection of the phrase “juvenile justice”. The proclamation of juvenile justice, which was supposed to guarantee their rights, has not become one of the institutions of the Russian judicial system. The reluctance of the authorities to create juvenile courts, to develop juvenile justice led to the fact that the concept of “juvenile” began to be perceived by society as something alien, inorganic.

It should be noted that the concept of “juvenile justice” is almost not used in international law, as this term is not used in the legal terminology of many developed democratic states. However, this does not mean that youth studies has outlived its usefulness. Most of the objective factors suggest otherwise. Let us highlight some of them.

1. Various political, economic, social, and cultural processes taking place in the country affect the entire society and all social strata. Freedom, which is today an attribute of the rule of law, requires certain legal regulation, carried out both by national legislation and by the norms of international law.

Children are also involved in these processes, moreover, as natural subjects of law. This requires expanding the scope of lawmaking, law enforcement practice of parents, of the state, and of the society to protect their rights—from the natural right to life to the positive right to personal freedom, individuality, protection of honor and dignity.

After the adoption and ratification of the Convention on the Rights of the Child, most states assumed obligations to protect the rights and interests of minors.

2. In modern state legal systems, the trend of specialization of law is widely used, which has not bypassed the juvenilizaton of law. Due to many objective conditions, subjective circumstances, the problems of protecting, ensuring the rights and interests of minors are becoming a kind of epicenter of such legal trends.

The main task of juvenile law today is the protection and maintenance of the rights and interests of minor children. This issue involves the restructuring of legal consciousness, legal relations, legal culture, primarily in relation to minors.

Juvenilization of the law also implies changes in the training, retraining, and advanced training of those involved in juvenile law-making and law enforcement activities. Certain knowledge in the youth studies is necessary for most of persons who defend the rights of minors. These are the parents and their substitute state structures, exercising their rights to protect minors.

3. In many states work has begun on the systematization of the legal norms that regulate the rights of minors and which are scattered across various branches of law, segments of legislation, materials of judicial practice.

In our country, such systematization of legislation, especially in the field of juvenile justice has not yet begun. It is drowning in endless and not always constructive discussions, and first of all, on issues of juvenile-procedural legislation. The National Strategy formed a certain concept of codification of the Russian legislation in the field of juvenile justice, outlined the directions of preparation of the relevant Federal Laws in the field of protection of the rights of minors.

4. Some authors, such as E.L. Voronova, Yu.A. Matskevich, suggest that:" in the context of the rejection of the "juvenile justice" concept by the mass public consciousness it will be the most convenient to use the term "juvenile legislation", "justice for minors in conflict with the law" instead [1].

We believe that this issue is not so simple and has deeper roots. Naturally, some part of society is quite critical of the concept of "juvenile justice". For one persons, this is taking on additional responsibility for the implementation of their parental responsibilities, for the second ones, it is the obligation to actively participate in increasing their legal, pedagogical, psychological competence, for the third ones, this is the realization of their political ambitions, for the fourth ones, this is a certain kind of PR, for the fifth ones and others, this is something blurred, new, exterior, alien to Russia.

In our state with its mentality, worldview, when new terms cause an acute reaction from society, the use of the concept of "juvenile justice" in Russian law, according to many legal scholars, has a negative meaning [10].

We adhere to the position reflected in the National Strategy on Action for Children for 2012–2017. Here the word "juvenileology" (youth studies) is used only once in the phrase "with specialists in juvenile technologies".

The content of the juvenility concept in this document is understood broadly and is associated, first of all, not with methods of coercion, but with methods of educational, preventive impact on a child and these children are not necessarily potential offenders. The provisions reflected in the Strategy have created the basis on which new sustainable ideas about the rights of children, about their interests in law, about the ways of realizing these rights not only by adults, but also by minor children themselves, are fundamentally built.

This approach to this problem, disavowing, obsoleting the juvenological terminology in a certain way, in fact fills with the sense of this term all the components of the legal, law enforcement system. Speaking about the terminology used in juvenology, it should be noted that it is used to meet the needs of the child. During the Soviet period, the social sciences paid great importance to the development of the theory of aspiring human needs. This is what lay at the basis of the causal relationship leading man to developed socialism and to communism. According to the teachings of the classics of Marxism, needs do not just develop, they are enriched with new content, aspiring the individual. The topic of the child's needs and their satisfaction in modern Russia has not lost its significance.

The needs of a minor, both positive and negative, arise from the moment of his birth and being in dialectical development, subsequently acquire real outlines in his interests and value attitudes. A large number of subjects of law are involved in the forming and implementation of these needs, including the family, parents, the state, certain institutions of civil society, and the children themselves. Here also questions arise that must be clearly answered. What needs of minors should be discussed? Who should determine their usefulness for the child, for the family, for the state, for civil society? What mechanisms and resources will be attracted for their implementation?

It is clear that it is necessary to talk about those needs that are constructive, and the criteria for the constructiveness of these needs are derived from the humanistic nature of the child. These needs must be recognized to a certain extent by the minor himself, his parents, and other adults who participate in their implementation. In real life, any need is recognized by each subject and is realized in its own way, from a subjective position. In such conditions, society, the state, forms certain moral and legal models of the most urgent needs. Within the framework of this model, a certain consensus is formed—a compromise between the child, his parents, other family members and institutions of upbringing, education, government bodies, and civil society entities. All these structures have physiological, psychological, material, spiritual, and other means that are capable of satisfying the constructive and humanistic needs of the child. This entire system of needs is the basis for their institutionalization and reflection in certain principles and norms of Russian law.

In international law, the system of universal needs of minors is fixed in the principles and norms set forth in the Convention on the Rights of the Child, as well as in a number of other international legal acts on the rights of the child.

Knowledge of the essential characteristics of the child's needs, the ways of their reflection in the public consciousness as the most significant axiological concepts allows state structures to implement them in practice and, accordingly, to consolidate this in principles, in the rule of law. At the present, state policy in the field of protecting the rights and interests of children is a priority in the Russian Federation. It is based on the following principles:

- legislative support of the rights of the child;
- support for the family as a whole;
- protection of the rights of minors;
- responsibility of officials and citizens for violations of children's rights;
- support of public organizations engaged in activities to protect the rights and legitimate interests of children [11].

As for the science of juvenile law, it is necessary to develop it, despite the fact that attention to this phenomenon is not as relevant today as it was ten years ago.

In conclusion, it can be noted that discussions on the relationship between the concept of juvenile law and the concept of law in the interests of children are promising and here there are original points of view that contribute to the development of juvenile law and juvenile justice. It seems to us that the subject field of law in the interests of children is much broader than the subject field of juvenile law. Juvenile law is a part of legal reality that regulates and protects the rights and interests of minor children, these are those moments in the legal life of a child under the age of 18, in which he participates as a person, citizen, individual, these are his actual rights, freedoms, and legal responsibility.

Thus, the interests of minors, like those of other people, are regulated through needs. Therefore, protecting the legitimate interests of children, the state legitimizes the needs that arise from these interests. The task of juvenile law is to create a legal framework in which the rights and interests of minors would be fully used.

References

1. Voronova EL, Matskevich YuD (2012) On the need to create a specialized judicial system for minors in the Russian Federation. Issues Juvenile Just 3:3–7
2. The Universal Declaration of Human Rights. Art. 1. In: Handbook on the implementation of the Convention on the Rights of the Child. Revised and rev, 3rd edn, UNICEF, 727 pp
3. Zaryzev AV, Malkov VD (2005) The concept of juvenile law, its subject, method and system. The Science of Juvenile Law. In: Saraeva AV, Malkov VD (eds) Juvenile Law: Textbook for high schools, Moscow, pp 23–24
4. Kamenskaya EV (2017) Modernization of law-making: main trends. Bull Moscow Witte Univ Ser 2: Legal Sci 1(10):3–8
5. The Constitution of the Russian Federation (adopted by popular vote on 12.12.1993). Article 2
6. The Constitution of the Russian Federation (adopted by popular vote on 12.12.1993). Article 17
7. The Concept of Juvenile Law in Modern Russia: monograph, VI Zhukov ed., Moscow, p 15
8. Lyublinsky PI (1908) Special Courts for Youth in North America and Western Europe/ P.I Lublinsky. SPb.: Senate typography, 82 pp
9. The National Strategy on Action for Children for 2012–2017. Approved by the decree of the President of the Russian Federation of June 1, 2012 N 761. In: Collection of legislation of the Russian Federation, No. 23, June 4, 2012

10. Nikolaeva YuV (2007) State Policy in the Field of Ensuring the Interests of Minors (Historical, Legal and Theoretical Problems). Monograph, pp 11–12
11. Nikolaeva YuV (2007) state policy in the field of ensuring the interests of minors (Historical, Legal and Theoretical Problems). Monograph, p 92
12. Rabets AM (2017) Juvenile Law of the Russian Federation: textbook, p 16
13. Sorochan VV (2015) Professional ethics, p 43
14. Tolstoy VS (2011) The concept of juvenile law. In: Tolstoy VS, Sumskoy DA (eds) Juvenile law: textbook for universities, p 19

Normative Legal Acts of Federal Executive Authorities

Olga N. Ordina

Abstract In the context of the implementation of transformations affecting many areas of public administration in the Russian Federation, special importance is attached to improving the rule-making of federal executive authorities. The paper states that the federal law regulating the general legal beginnings of their activities has not yet been adopted, and there is no legislative basis for the creation of regulations by these bodies. The current federal administrative law does not enshrine the general legal concept of the legal act of the federal executive branch (both normative and individual) and its features. At present, the issue of the types of regulations of the federal executive authorities, their name and the official form of publication (acceptance) has not been settled. This paper is an attempt at a comprehensive scientific study of normative legal acts of federal executive authorities, namely, a theoretical understanding of the concept, features, types, forms of these acts, identifying the reasons for existing shortcomings in the legislation and making proposals for its improvement.

Keywords Regulatory legal acts · Federal executive authorities · Management acts · Federal ministries · Federal services · Federal agencies

1 Introduction

The object of our investigation is the legal acts of the federal executive branch, including officials acting on behalf of those bodies. Their legal nature and place in the legal system have not been sufficiently investigated.

The content of government acts is crucially determined by the scope and nature of the competence of the Government of the Russian Federation. The Government,

O. N. Ordina (✉)
Moscow Witte University, 12, 2nd Kozhukhovsky Pas., Bldg. 1, Moscow 115432, Russia
e-mail: ordinolga@yandex.ru

Russian University of Transport RUT (MIIT), 22/2, Chasova Street, Bldg. 1, Moscow 125315, Russia

A. V. Semenov and I. A. Sokolov (eds.), *Sustainable Development: Society, Ecology, Economy*, Earth and Environmental Sciences Library,
https://doi.org/10.1007/978-3-030-73110-6_15

which exercises the executive power of Russia, under the current Constitution of the Russian Federation, has the main function in organizing the general and current leadership of economic, socio-cultural and administrative and political development. The regulatory legal acts of the Russian Government have a powerful law-making potential.

The regulations of federal ministries, federal services and federal agencies are called official regulations. These acts were studied by both Soviet and Russian scientists. These problems were developed in the works of Nikolaeva [30], Nozdrachev [31], Pigolkin [32], Tikhomirov [36] and others.

A similar theme has been developed in the works of foreign scientists such as Akehurst [40], Debbach [41], Danilenko [42], Gellhorn [43] and others.

The system and structure of the federal executive branch are enshrined in decrees of the President of the Russian Federation on March 9, 2004, №. 314 [5], May 15, 2018, № 215 [17] and others.

In its ruling of January 27, 1999 № 2-P [19], the Constitutional Court of Russia stated that on the issues of the federal executive branch, not settled by the legislator, the President of Russia has the right to issue decrees that should not be contrary to the Russian Constitution and federal laws.

In his decree № 314, the head of state indicated that federal executive bodies do not have the right to adopt normative legal acts, except for the cases established by decrees of the President of Russia or decrees of the Government of Russia. For example, paragraph 8 of the Decree of the President of the Russian Federation № 649 of May 20, 2004 "On Issues Related to Structure of Federal Executive Bodies" [6] indicates that various departments headed by the Russian Government have the right to adopt regulatory legal acts. It should be noted that federal services and federal agencies make extensive use of the exceptions granted to them, thus it can be stated that they carry out normative legal regulation in the established area of activity along with the federal ministries.

Institutional regulations can be both interagency and over-agency and regulated by individuals and entities.

2 Analysis of the Legal Nature of the Regulations of the Federal Executive Branch of Russia

Let us analyze the legal nature of government and departmental regulatory administrative legal acts by analyzing scientific concepts that exist in Russian science.

Regulatory acts of the federal executive branch of Russia are a type of administrative and legal acts, which are also called acts of state administration, legal acts of administration, administrative acts, etc. Depending on legal content, administrative acts are divided into normative and individual acts. Normative administrative legal acts are sources of administrative law.

In the legal literature of the pre-Soviet period, considerable attention was paid to the normative administrative law act. Among the forms of law G.F. Shershenevich singled out an administrative decree from. In his opinion, the signs of an administrative decree are: its issuance by an authorized governing body; non-contradiction of the decree and the law; notoriousness of the decree, i.e. bringing it to the attention of the population [38, p. 78-83]. In the works of N.M.Korkunov, A.I. Yelistratov and other scholars, the administrative act was considered as a means of regulation and protection of public law relations. In their opinion, an administrative act is a command of the state administration bodies, which is adopted within the limits of discretion, the scope of which is determined by law. A.I. Yelistratov wrote: "A normative act formulates legal norms, i.e., general rules dictated by state power and applicable to all cases of life covered by their meaning" [27, p. 132]. N.M. Korkunov noted: "Unlike legislation, acts of government activity do not represent the unity of form." "Legislation is, as it were, an element of constancy, sustainability; government is an element of movement" [29, p. 220].

In the Soviet period, normative legal acts of government were considered as acts of will of state bodies, which are carried out by them in compliance with and on the basis of laws containing the norms of law [26]. The following features were distinguished: legality, authority (obligation), stability and relevance [34, pp. 197–204].

In the Russian legal literature, depending on the understanding of public administration in a "narrow" or "broad" sense, the following points of view have developed about the concept of a regulatory administrative act: some scholars (for example, Yu. A. Tikhomirov [35, p. 132] believe that that a regulatory legal act of public administration is a legal form of executive and administrative activity carried out by all state bodies (both legislative and executive); others (for example, A. A. Savostin, Yu. N. Starilov, M. V. Galkina and P. I. Kononov) define a regulatory legal act of public administration as a legal act issued by public administration bodies [33, pp. 53, 54; 23, p. 321].

The institutional act of the federal executive branch is considered by various authors in different ways: first, as an official written document (Zhchevakin [28, p. 4], Chernikov [37, p. 4]); second, as subordinate legislation (Shmakova [39, p. 95]); third, as an element of the mechanism of legal regulation (Arzamasov [22, p. 203]), fourth, as a legally imperious will (Borovikova [24, p.21]).

In our opinion, government and departmental regulatory administrative legal acts have the following features.

1. These acts are one of the types of regulatory legal acts. Therefore, these acts have their characteristics: statutory, normative, formally defined, confirmative). Currently, the current legislation of Russia lacks a general concept of a normative legal act. The legislation also does not define the concepts of a normative administrative-legal act and a normative legal act of the executive authorities of the federal level.

S. V. Boshno considers the generally accepted definition of a normative legal act as a written document, which is adopted by an authorized subject of law, has an official character, is obligatory, and contains legal norms [25]. The federal law of

1996 "On Regulatory Legal Acts in the Russian Federation" draft also contained a definition of the concept of an act of normative content [20].

Since there is no legislative consolidation of the definition of the "normative legal act" concept, there are many questions arise when determining the normative administrative legal acts of the federal level.

2. These acts are subordinate normative legal acts.

One of the functions of the Government of Russia is to organize the execution of federal laws, which determines the nature of its acts, namely, their "subordination", i.e. their adoption on the basis of and in pursuance of laws and decrees of the President of Russia.

State and public interests require the Russian government and other executive authorities to act within the law. In reality, the connection between the norms of laws and regulations is often violated, the latter are slowly changed and canceled. Currently, the task of increasing the significance of the law as the basis for making decisions by the Government of Russia is still urgent.

A number of government acts are adopted "on the basis of the law", some "in accordance with the law", and some of these acts "in pursuance of them".

In the first two cases ("on the basis of the law" and "in accordance with the law"), guided by the sense and general principles of one or more legislative acts regulating the range of public relations, the Government of the Russian Federation is more free in choosing the means and methods for solving the problem. When issuing a management decision "in pursuance", the Government of Russia is more closely connected with the goals and content of the legislative act, therefore, in this case, there is a closer structural and logical connection between the acts. Since such a government act follows the legislative act, it contains a large number of reproduced and subordinate statutes.

Other executive bodies of the federal level adopt administrative acts that also have a feature of subordination, i.e. adopted on the basis of and pursuant to laws, acts of the President of Russia and the Government of Russia and cannot contradict them.

3. These acts are issued or adopted within the competence of this or that body. Due to this, they are a way of realizing the competence of the Russian Government and departments. The issues of the jurisdiction of each department are established in the regulations on these bodies.
4. These acts, performing the function of implementing normative legal acts of higher state bodies, can introduce additional details into the mechanism of action of the norms being implemented, thereby affecting the effectiveness of legal regulations.
5. These acts are subdivided into acts of non-departmental actions that are beyond a certain body and are mandatory for all entities not subordinate to this body, and departmental actions, which are aimed at regulating intra-departmental relations within a certain body.
6. The legal acts of the federal executive branch differ in form and content, due to the nature of their powers, the specificity of their competence and the nature of the tasks they are tasked with.

7. These acts contain the norms of administrative law. However, they can enshrine, change or abolish not only the norms of administrative law, but also the norms of other branches of law, in particular, civil, financial, environmental, housing. Administrative legal acts may contain the norms of these branches of law and not contain the norms of administrative law [21, p. 342]. In other words, in terms of content, administrative and legal acts can be not only administrative, but also civil, land, environmental, financial, etc. and act as sources of relevant branches of law.

3 Classification of Regulations Right-Wing Acts of the Federal Executive Branch

The legal nature of the regulations is becoming more understandable by classifying these acts.

The criteria for dividing these acts may be the following: (1) legal force; ((2) scope of action; (3) subjects of acceptance; (4) the territory of action; (6) normative content; (7) functional role; (8) form, etc.

Depending on **the legal force**, normative administrative legal acts of the federal level are divided into the following types: (1) decrees of the Government of Russia; (2) regulatory legal acts of federal ministries, federal agencies of federal services.

Depending on **the scope of application**, regulatory administrative acts of the federal level are of the following types:

(1) intradepartmental normative legal acts regulating the internal activities of the system of bodies, institutions and organizations of a certain federal executive body (for example, order of the Ministry of Internal Affairs of Russia №. 1 [16]);
(2) supra-departmental normative legal acts that extend their effect to individuals and legal entities that are not part of the system of the body that adopts (issues) these acts (for example, order of the Ministry of Internal Affairs of Russia No. 623 [18]);
(3) interdepartmental normative legal acts adopted (issued) by the heads of several federal ministries, federal services, federal agencies, and regulating the interaction of these bodies.

Depending on the entities that have the right to adopt (issue) regulatory administrative and legal acts of the federal level, the following types are highlighted:

(1) administrative and legal acts adopted (issued) independently by any federal executive body (for example, order of the Ministry of Culture of Russia № 558 [15]);
(2) regulatory legal acts adopted (issued) by several federal ministries, federal services, federal agencies (for example, order of the Ministry of Internal Affairs of Russia № 227 and other bodies [7]).

Depending on the **territorial scope of application**, federal administrative and legal acts are divided into: (1) acts extending their effect to the entire territory of Russia; (2) acts extending their effect to the territory of a certain constituent entity of the Russian Federation; (3) acts extending to the territory of a particular municipality.

Depending on the normative content, it is possible to divide the regulatory administrative and legal acts of the federal level into:

(1) regulatory legal acts that contain exclusively the norms of administrative law (for example, the Traffic Rules [1]);
(2) regulatory legal acts that contain both the norms of administrative and other branches of law.

Depending on **the functional role**, the following types of these acts can be distinguished:

(1) acts in connection with the approval of regulatory administrative and legal acts of the federal level (for example, order of the Ministry of Transport of Russia of June 26, 2008 № 94 [8]);
(2) acts on amendments to regulatory administrative and legal acts of the federal level (for example, Resolution of the Government of Russia of December 29, 2008 № 1048 [11]);
(3) acts on the introduction of amendments and additions to the normative administrative and legal acts of the federal level (for example, the order of the Ministry of Agriculture and Food of Russia of March 9, 1999 № [3]);
(4) acts canceling other normative administrative and legal acts of the federal level (for example, the decree of the Government of Russia of March 20, 2006 № 146 [8]).

Depending on **the official-documentary form** of expression, these acts are the following types:

(1) resolutions (adopted by the Government of Russia);
(2) orders (mostly adopted by the heads of federal ministries);
(3) rules (approved by orders of federal ministries and decrees of the Russian Government);
(4) provisions (approved by acts of federal ministries and the Government of Russia);
(5) charters (approved by acts of federal ministries). For example, the "Disciplinary Charter of the Customs Service of the Russian Federation" [2];
(6) guidelines (approved by acts of federal ministries). For example, "The Federation's Technical Oversight Board of the State Road Safety Inspectorate of the Russian Ministry of the Interior" [4];
(7) instructions (accepted by the federal executive authorities). For example, "Instruction on compulsory public insurance for the life and health of employees of institutions and bodies of the penal system" [9];
(8) administrative and other regulations (approved by orders of federal executive bodies). For example, "Administrative Regulations of the Federal Agency

for the Management of Public Property on the provision of public services "Exercise in due course the issuance of statements from the registry of federal property" " [13];

(9) orders (issued by the federal executive authorities), etc. For example,"The interaction of border authorities and customs authorities in the conduct of inspection operations for vehicles, goods and goods transported across the state border of the Russian Federation in road and sea checkpoints across the state border of the Russian Federation" [12].

In our opinion, these are the most important grounds for the classification of administrative and legal acts of the Russian Government, federal ministries, federal services, federal agencies.

Based on the foregoing, the following conclusion can be drawn. The system of regulatory administrative and legal acts of the Government of Russia, federal ministries, federal services, federal agencies is a complex phenomenon. We have considered the problems of the legal nature and classification of these acts, which require close attention and resolution.

4 Conclusion

Currently, there is no legislative definition of a regulatory administrative and legal act of the Russian Government, federal ministry, federal service, federal agency, and this circumstance negatively affects the activities of both these bodies and other state bodies. Therefore, it is required to enshrine the specified definition in legislative acts.

A gap in legislation, both in terms of defining various forms of regulatory administrative and legal acts of the federal level, and in terms of consolidating the criteria for their delimitation, the dynamism and variety of issues solved with the help of these acts lead to disorder in the rule-making of the Russian Government, federal ministries, federal services and federal agencies. Each normative legal act of these bodies should have a legislatively enshrined form that largely determines its individual characteristics and content. The development and adoption of the federal law "On legal acts of federal executive bodies in the Russian Federation" will contribute to the establishment of uniformity of regulatory administrative and legal acts of the federal level outside their departmental affiliation.

References

1. On the rules of the road: the resolution of the Russian Government of October 23, 1993. № 1090. Collected Acts of the President and the Government of the Russian Federation. 1993. № 47. Art. 4531

2. On the approval of the Disciplinary Charter of the Customs Service of the Russian Federation: decree of the President of the Russian Federation of November 16, 1998. № 1396. Collected Acts of the Russian Federation. 1998. № 47. Art. 5742
3. On amending and supplementing the order of the Russian Ministry of Agriculture on September 23, 1997 № 404 « On the approval of the Rules for the preparation of regulations of the federal executive authorities and their state registration » the order of the Ministry of Agriculture of the Russian Federation on March 9, 1999 № 100. The document was not published. Consultant Plus
4. On the approval of regulations governing the activities of the State Road Safety Inspectorate of the Ministry of the Interior of the Russian Federation on technical supervision: the order of the Ministry of Internal Affairs of the Russian Federation of December 7, 2000. № 1240 (Ed. from February 19, 2007). Bulletin of the regulations of the federal executive authorities. 2001. № 7
5. On the system and structure of the federal executive branch: decree of the President of the Russian Federation of March 9, 2004. № 314/ Collected Acts of Russian Federation. 2004. № 11. Art. 945
6. Issues of the structure of the federal executive branch: decree of the President of the Russian Federation of May 20, 2004. № 649/ Collected Acts of Russian Federation. 2004. № 21. Art. 2023
7. On the creation of the Interdepartmental Commission on the ordering of entry and stay in the territory of the Russian Federation of foreign citizens and stateless persons: the order of the Russian Foreign Ministry № 3769, the Ministry of Internal Affairs of Russia № 227, the Ministry of Health and Social Development of the Russian Federation № 244, the FSS of the Russian Federation № 183, the RFI № 20 of March 30, 2005. Bulletin of Federal Agency Regulations. 2005. № 21
8. On the abolition of some decisions of the Government of the Russian Federation: the resolution of the Government of the Russian Federation of March 20, 2006. № 146. Collected Acts of Russian Federation. 2006. №13. Art. 1398
9. On the approval of the Instruction on the conduct of compulsory state insurance for the life and health of employees of institutions and bodies of the penal-executive system: the order of the Ministry of Justice of the Russian Federation of April 13, 2006. № 114. Bulletin of the regulations of the federal executive authorities. 2006. № 20
10. On the approval of the Order of action of participants in the transport process on the railway in case of natural and man-made emergencies: the order of the Ministry of Transport of the Russian Federation of June 26, 2008 № 94. Bulletin of Federal Agency Regulations. 2008. № 30
11. On amending the « Rules for the preparation of regulations of the federal executive authorities and their state registration » : the Russian government's resolution of December 29, 2008. № 1048. Collected Acts of Russian Federation. 2009. № 2. Art. 240
12. On the approval of the Order of Interaction between border authorities and customs authorities in the conduct of inspection operations in respect of vehicles, goods and goods moved across the state border of the Russian Federation at road and sea checkpoints across the state border of the Russian Federation: the order of the FEDERAL Security Service of the Russian Federation No. 49, the Federal Security Service of the Russian Federation № 167 of February 11, 2009. Customs Bulletin. 2009. № 6
13. On the approval of the Administrative Regulations of the Federal Agency for the Management of State Property on the provision of public services « Implementation in the prescribed manner of issuing extracts from the register of federal property » : the order of the Ministry of Economic Development of the Russian Federation of June 1, 2009 № 201. Bulletin of Federal Agency Regulations. 2009. № 34
14. On the approval of the Regulation on the specifics of providing technical conditions, the determination of fees for technological accession and features of technological accession to the facilities of the electricity grid of electric power-receiving devices of consumers of electric energy in the city-resort of Sochi municipal entity during the organization and conduct of the

XXII Olympic Winter Games and the XI Paralympic Winter Games in 2014, and the amendment smending some acts of the Government of the Russian Federation Federation: Russian Government Resolution of May 31, 2010 № 341. Collected Acts of Russian Federation. 2010. № 21. Art. 2607
15. On the approval of the « List of typical management archival documents formed in the course of the activities of state bodies, local governments and organizations, specifying the terms of storage » : the order of the Russian Ministry of Culture of August 25, 2010 № 558). Bulletin of Federal Agency Regulations. 2010. № 38. 20.09.2010
16. On the approval of the Instruction on the procedure of assigning qualification titles to the employees of the internal affairs bodies of the Russian Federation: the order of the Ministry of Internal Affairs of the Russian Federation of January 10, 2012. № 1/ Russian Newspaper. 2012. (№ 76)
17. On the structure of the federal executive branch: decree of the President of the Russian Federation on May 15, 2018. № 215/ Collected Acts of Russian Federation. 2018. № Art. 2981
18. On the approval of the Instructions on the organization of the activities of the Ministry of the Interior of the Russian Federation and its territorial bodies in the consideration of applications on citizenship issues of the Russian Federation and decisions on them: the order of the Russian Ministry of the Interior of September 16, 2019 № 623/ The official Internet portal of legal information http://www.pravo.gov.ru. Accessed 12 Nov 2019
19. On the case of interpretation of Articles 71 (paragraph « D (Г) »), 76 (part 1) and 112 (part 1) of the Constitution of the Russian Federation: the ruling of the Constitutional Court of the Russian Federation of January 27, 1999 № 2-P. Collected Acts of Russian Federation. 1999. № 6. Art. 866
20. Draft Federal Law "On Regulatory Legal Acts of the Russian Federation". Journal of Russian Law. - 1997. - № 10. - P. 157
21. Kononov PI (ed) Administrative Law of Russia: textbook for higher education institutions. Unity-Dana
22. Arzamasov YG (2004) Institutional Rule-Making Process in the Russian Federation: dis.... Dr. of Law Sciences
23. Bahrah DN, Rossiynskiy BV, Starilov JN (2004) Administrative Law: a Textbook for Universities
24. Borovikova YP (2004) Regulatory Legal Acts of the Federal Executive Branch of the Russian Federation: autoref. Dis.... cand. of Law Sciences
25. Boshno SV (2009) Regulatory Act: Development of Features in Law Enforcement Practice. Arbitrazh and Civil Procedure, № 4
26. Vasilyev RF (1970) Legal Acts of Management
27. Yelistratov AI (1929) Administrative Law
28. Zhevakin SN (1998) Departmental Regulations in the Russian Federation. - Voronezh
29. Korkunov NM (1913) State Law. - St. Petersburg
30. NikolaevMN (1975) Regulations of the Ministries and Departments of the USSR
31. Nozdrachev AF (1968) Regulations of the Ministries and Departments of the USSR: dis.... cand. Of Law Sciences
32. Pigolkin AS (1972) Theoretical Problems of Law-Making in the USSR: dis.... Dr. of Law Sciences
33. Savostyn A (2002) Characteristics of the Administrative Act as a Form of Expression of the Method of Regulating Public Relations. Jurist,- № 2
34. Kozlov YM (1962) Soviet Administrative Law: textbook
35. Tikhomirov YA (2005) Administrative Law and Process: Full Course/ Y.A. Tikhomirov. M.: publ. of Tikhomirov MYu
36. Tikhomirov YA, Kotelevskaya IV (1999) Legal Acts: manual
37. Chernikov VV (1996) Normative Acts of the Internal Affairs Bodies in the System of Legal Acts of Management: lecture

38. Shershenevich GF (19966) General Theory of Law: textbook./ Manual (according to the edition of 1910–1912), vol 2
39. Shmakova NS (2006) departmental law-making: concept and forms: dis…. Cand. of Law sciences
40. Akehurst M (1987) A Modern Introduction to International Law/ M. Akehurst
41. Debbach Ch (1971) Science administrative/ Ch. Debbach. – Dalloz
42. Danilenko GM (1993) Law – Making in the International Community/ G. M. Danilenko, Dordrecht
43. Gellhorn E (1981) Administrative Law and Process/ E. Gellhorn, B. B. Boyer, 2nd ed. 3rd reprint

The Concept and the Main Categories of Legal Pedagogy

Ilya I. Aminov

Abstract A general characteristic is given of a relatively new branch of scientific knowledge—legal pedagogy, which arose at the junction of jurisprudence (legal sciences) and pedagogy. In legal pedagogy, the joint achievements of both sciences (their methodologies, theories, categorical apparatus, practical developments) find their fullest expression. As a branch of pedagogy with a clearly predominant pedagogical approach in highlighting its own problems, it is aimed to solve many problems similar to jurisprudence: based on the demands and needs of society, to prepare lawyers for professional activities related to the development and implementation of legal norms; ensuring the rule of law and order, conducting scientific research, education and training.

Keywords Pedagogy and jurisprudence · Categories of legal pedagogy · Elements of the pedagogical system · Law enforcement · Law application · Rule-making activities · Private pedagogical categories · Patterns of legal pedagogy

1 Introduction

In the era of total and large-scale digitalization, it turns out that there are still spheres that are not subject to essential virtual impact. Among such, virtually untouched, phenomena is a legal pedagogy. Of course, information technological processes may, to one degree or another, digitalize individual organizational and functional pedagogical operations, however, in general, legal pedagogy remains neutral to digitalization, while remaining focused on live, physical human communication.

Legal pedagogy is a field of scientific knowledge that develops at the junction of jurisprudence and pedagogy, a type of professional pedagogy. Its purpose is the maximum possible social protection of the interests and rights of every person, assistance to the functioning and development of civil society.

I. I. Aminov (✉)
MGIMO (University) Under the Ministry for Foreign Affairs of Russia, Prospekt Vernadskogo, 76, Moscow 119454, Russia
e-mail: aminovii@mail.ru

A. V. Semenov and I. A. Sokolov (eds.), *Sustainable Development: Society, Ecology, Economy*, Earth and Environmental Sciences Library,
https://doi.org/10.1007/978-3-030-73110-6_16

Civilized forms of human behavior that meet the ideals and norms of modern society are unattainable without an appropriate level of education, upbringing, training level and maturity of citizens. The level of legal culture of society, morality, legality also depends on these reasons, however no less significant here is the continuous pedagogical work with the population with the powerful and comprehensive support of carefully trained, pedagogically competent lawyers. Being engaged in solving specific problems related to legal education, professional legal training, professional development, legal propaganda, providing pedagogically expedient decisions and actions in the legal field, legal pedagogy is aimed to ensure the scientific level of this work. Such an enormous coverage of social significance testifies to the breadth and constructiveness of the use of its data in the training of legal personnel, in solving professional problems facing officials and specialists.

> The above allows us to define legal pedagogy as an applied branch of knowledge, aimed to solve specific methodological issues that are directly related to the professionalism of lawyers in different areas of activity and the real ways they achieve the best results in professional and personal growth [1].

The possibilities of using the data of legal pedagogy in ensuring the rights and freedoms of citizens should not be underestimated. The importance of pedagogical recommendations related to these tasks is determined not only by their closest connection with each other. At present, the other side of reality is not sufficiently understood by practicing lawyers: practically all specialists in the field of law, officials of the law enforcement and law application system of the state are related to pedagogical factors that influence the strengthening of law and order. For some, it is presented as a kind of professional and pedagogical activity (teaching staff of law schools of the Ministry of Science and Higher Education, educational institutions of the Ministry of Justice, the Investigative Committee, the Ministry of Internal Affairs, the Federal Penitentiary Service, the Federal Customs Service of the Russian Federation; employees providing work with personnel, professional training in practical bodies); in others, it is characterized by the presence of clearly expressed legal and educational functions (local district police chief, PIS employees, STSI, police dispatch centers of local internal affairs bodies, as well as those who are directly related to the delinquent behavior of minors); in others, it is distinguished by the presence of the legal educational and educational influences of the very personality of a professional lawyer on citizens, its actions, style of behavior, methods used (employees of the judicial system, representatives of the lawyer community, heads of law enforcement and law application agencies, employees of the prosecutor's office, the Investigative Committee, the Federal Customs Service, Federal Bailiff Service, etc.).

The results of the activities of each of the listed employees and officials, obtained by pedagogically responsible and competent efforts, contribute to the optimization of professional (service) indicators, have a positive impact on the prevention of administrative offenses and the prevention of crime. Therefore, at present, it can be argued without exaggeration that legal pedagogy is outwardly inconspicuous, however, persistently, relying on the powerful potential of the listed departmental and law enforcement agencies, with a "sense of law and rightness" introduces topical scientific developments that, being in demand in state-legal construction, able to affect the minds of lawyers and citizens more than hundreds of universities.

Thus, legal pedagogy accompanies all areas of a lawyer's activity—rule-making, law enforcement, law application, expert consulting, organizational and managerial, research, and pedagogical itself. As a branch of applied knowledge, this discipline is also valuable since it is constantly enriched through interaction with related branches of scientific knowledge, generalizing them in a system of means, forms and methods necessary for lawyers for professional and personal growth. Thus, from pedagogical psychology, it perceived the methods of teaching and upbringing that increase the efficiency of solving educational problems; from labor psychology—provisions covering the activities of law enforcement and law application officials; from the psychology of creativity—information about the solution of problems, similar to those that are solved in the framework of lawmaking; from legal psychology—information about the means (techniques, methods) of educational influence on the suspect, accused, witness in situations of conflict interaction; from pedagogy and psychology of higher education, correctional pedagogy—recommendations on moral and legal education, teaching student youth, prevention of alcoholism, drug addiction and other addictions in this still unformed socially maturing environment. Since the preliminary investigation, trial, accused's defence, no less than other acts committed by a lawyer, are carried out within the framework of communication, legal pedagogy is closely linked with social psychology and professional ethics. Achievements of modern social pedagogy are of particular relevance for legal pedagogy. Thus, along with the implementation of general pedagogical laws in the educational process of educational institutions that train lawyers, social pedagogy focuses on the study of the social impact on personality formation, develops a system of measures to optimize this process. At the same time, the value of social pedagogy for lawyers also lies in the fact that it concentrates the areas of scientific research not only on the issues of education of different age groups and categories of citizens, carried out in organizations specially created for this, but also in those where education is not the main function.

All of the above mentioned areas indicate that legal regulation is objectively conditioned by social and socio-pedagogical laws. Therefore, only by combining legal knowledge with knowledge of the above mentioned branches of pedagogy, psychology, ethics, etc., students will be able to become truly competent professionals in the field of law, they will be able to work with citizens and special contingent.

Based on the understanding of the subject of legal pedagogy, the experience of related fields of scientific knowledge and the results of the research carried out, at present we can talk about certain groups of its conceptual and categorical apparatus and, first of all, about the specifics of such ***basic pedagogical categories*** as education, upbringing, training, pedagogical activities, etc., which have received the "right of citizenship" in the methodology of legal pedagogy.

Despite the constant updating and evolution of the content of the listed general pedagogical concepts, legal pedagogy meets social expectations, i.e., it gets with the times. Thus, the Federal Law 273-FZ of December 29, 2012 "On Education in the

Russian Federation" introduced clear definitions of such frequently used categories as "education", "upbringing", "training" (Art. 2), which are important both for the general and for legal pedagogy.

Upbringing means the process of purposeful impact on the spiritual and physical development of a person in order to prepare it for social, industrial, cultural activities. However, until the beginning of the eighteenth century, the verb "upbring" had only a physical meaning: to feed, to nurture. In the eighteenth century this word began to be understood differently—as a desire to develop a person mentally and morally [2, p. 55]. At present, the verb "upbring" has entered the "third phase of its interpretation" and means that what is closest in the sense of legal pedagogy—"to impact the formation of a person's value system" [3, p. 34].

In pedagogy, the question of differentiating the concepts of upbringing and socialization is often discussed. Despite the fact that socialization is more often viewed in the context of uncontrolled upbringing, legal pedagogy is currently solving the problem of creating such a system of upbringing, thanks to which a person, socializing (re-socializing), would strive for self-improvement, elimination of personality defects.

A number of scientists associate upbringing only with the moral formation of the individual, and everything else for them is just communication. This position has areas of common interest with legal pedagogy, since upbringing by morality leads to a certain conceptually formed system of moral norms, all recognized patterns of behavior, contributes to the translation of moral knowledge into moral convictions of the individual, develops the skill of moral interpretation of specific, including legally significant situations.

> We live and act, we born and die, rejoice and suffer, surrounded by the" air of law",—wrote the Russian sociologist and educator P. A. Sorokin (1889–1968),—"in this sense, law penetrates into all the pores of social life, like an ethereal spirit, it accompanies us everywhere and controls us. Thanks to it, we are indignant at the sight of this or that meanness, thanks to it we are indignant at the lack of rights and rush into the struggle for the right, it arouses in us admiration for "high acts of fulfillment of duty", it gives us an indication of how to act in one case or another, it gives us confidence in the use of our powers—all these various actions, feelings, experiences caused by our legal convictions, agreeing with them and realizing them, are nothing more than forms of manifestation and implementation of rights or legal convictions [4, p. 29].

Education is a socio-cultural phenomenon, an institution for the life support of society in specific historical conditions, a socially significant benefit carried out in the interests of a person, family, society and state. Therefore, the objective prerequisite for the emergence of educational relations is not only personal social welfare, but also social necessity.

In the Christian tradition, the origin of the word "education" is understood from the words "image" as an iconographic image of the Holy Face of God and "sculpturing", in other words, the creation of something ideal, and therefore education is the path of human self-development in his approach to God. There are approaches where the process of acquiring knowledge is considered as self-sufficient, not requiring separate educational work. Nevertheless, in the pedagogical process, educational influences are always present, since education is realized in teaching.

Based on the tradition closest to legal pedagogy, a different meaning is put into education: a person chooses an image for himself, an ideal that it strives to be equal to and which it approaches in the process of education and personal growth.

It is notable that the first People's Commissar of Education of the Soviet Republic, A. V. Lunacharsky (1875–1933), wrote about the difference between training and education:

> … education is a creative process. Throughout its life, a person's personality is "formed", expands, enriches, strengthens and improves [5, p. 366].

Among the functions that education performs, teachers highlight: the entry of a person into the world of science and culture; ensuring the continuity of generations; broadcasting samples of human activity and behavior; meeting educational needs and interests.

According to Part 2 of Art. 10 of the Law on Education in the Russian Federation, education is divided into general, vocational, additional and vocational training. These directions ensure the possibility of realizing a person's right to education throughout his life.

Training is a process of stimulating and managing the external and internal activity of an individual, as a result of which the development of human experience takes place [6, p. 177]. Training involves the pedagogical activity of the teacher and the educational activity of the student. Since the activity of a teacher in teaching is called teaching, and the activity of a student is called learning, teaching is also defined as follows: it is teaching and learning taken in unity. At the same time, teaching differs from learning in different accents: teaching focuses on the input of the pedagogical process, that is, on what is given to the student—knowledge, skills, abilities. Learning focuses on output, that is, on what the learner has received. At the same time, the acquired is assessed by him not so much from the point of view of its content or interest as from the point of view of usefulness for present or future professional activity.

Development is the process of personality formation under the influence of external and internal, controlled and uncontrolled social and natural factors. For a thorough study of this process, science follows the path of differentiating its components, highlighting in it the development of physical, mental, spiritual, social, etc. Pedagogy, including legal, is focused on the problem of the spiritual development of the individual in conjunction with all other components [7, pp. 196–197].

The intensification of personality development occurs in the process of activity (including educational). This function is more successfully implemented in a number of special pedagogical technologies—a certain set of methods in the pedagogical process that pursue precisely the goals of personality development—the most complex process of objective reality.

Pedagogical activity is a special type of professional activity aimed at realizing the goals of education. The goals of education are not only set from the outside, but also arise within the pedagogical process. The English sociologist E. Giddens (born in 1938)—the author of the cognitive model of socialization, claims that the student is not a passive being, amenable to "cultural programming", on the contrary, it is his needs that guide the teacher in acquiring certain knowledge or building a pedagogical process.

Let us also name the systemic categories that characterize ***the most important elements of legal and pedagogical reality***. These, first of all, should include:

(a) a pedagogical approach to law, which involves the analysis of all issues of law enforcement and law application practice on the basis of scientific and pedagogical knowledge, the search for answers to them using pedagogical methods, means, techniques and technologies;
(b) a pedagogical system for ensuring law enforcement, law application, rule-making and other activities, which is an integral unity of state and public pedagogical institutions, the activities of which are consolidated by the common goals and objectives of legal training, education and upbringing;
(c) legal education, which is professional, considered as a special pedagogical system, process and result necessary for employees of law enforcement, law application agencies, legal services of enterprises (organizations), etc., that meets the requirements of the Federal State Educational Standard. Since in the strategic plan there is no alternative to the most developed, i.e., scientific knowledge, education in higher education institutions is based on the study of sciences. For teachers there is no alternative to scientific pedagogy, and for lawyers—the science of jurisprudence [8, p. five];
(d) legal training, which is a purposeful, systematic and organized process of bilateral activities to equip lawyers and citizens with legal knowledge, skills and abilities necessary for them to successfully perform professional (official) tasks and lawful behavior. Legal education is closely related to legal upbringing. They are divided according to the sphere of influence: training affects, as a rule, on the cognitive-rational side of consciousness, upbringing—on the emotional-volitional, value, worldview, spiritual side [9, p. 18];
(e) legal development, considered as a process of progressive changes occurring in the personal and activity characteristics of lawyers and citizens and expressed in an increase in their aspiration for professionalism and lawful behavior. Legal personality development is an important condition for success in life and a high position in society. It is expressed in legal convictions (ideological positions and attitudes); legal relationships (interrelations with life and work); legal habits (established ways of civilized behavior) [10, p. 250]. The great responsibility and social significance of the tasks solved by the lawyer, the exceptional diversity and complexity of the social functions performed by him, make it necessary for him to have development in other areas of humanitarian knowledge, including pedagogy, psychology, ethics;
(f) legal enlightenment, which involves increasing the level of legal literacy and the development of legal awareness of citizens, providing upbringing impact in order to prevent the commission of offenses or antisocial behavior. If legal upbringing is the most important of the methods of instilling legal culture in

citizens by state bodies, institutions, organizations, then legal enlightenment is a more capacious legal and pedagogical category. It is aimed at raising the level of legal awareness of citizens on the part of not so much authorized subjects, but rather using public resources and potential;

(g) vocational training, which is a special pedagogical system, implemented, as a rule, directly in a practical body (institution) and solving the problem of forming, improving and maintaining the professional competence of the staff. At the same time, relations arising in connection with vocational training, retraining and advanced training are directly related to labor (service) relations.

Thus, legal and pedagogical reality as a whole is not only and not so much teaching students in criminal, civil law and other industry disciplines, it is a social reality that is directly related to legality, combating offenses and crime, fostering legal culture and law-abiding behavior, teaching highly skilled actions in difficult, often extreme situations of professional activity.

The high social responsibility assigned to legal pedagogy determines the content of the research tasks formulated by it. It becomes important for it not only the development of problems of legal regulation, prevention and fight against crime, i.e. causes and consequences that determine the state of law and order. It is included in the study of topical issues of the development of the legal position of citizens, their legal activity—the most important, complex and underinvestigated phenomena of our time.

References

1. Stolyarenko AM (1999) Legal pedagogy in the system of the ministry of internal affairs: methodology, theory, practice: Author. dis. … Dr. Ped. Sciences. Military University of Russia, Moscow
2. Vinogradov VV (1994) History of words, Shvedova NYu (ed) Tolk
3. Krasnova TV (2016) Education as a unique legal phenomenon and problems of its modern legal definition. Russian Laws: Experience, Analysis, Practice, no 9
4. Sorokin PA (1919) Elementary textbook of the theory of law in connection with the theory of the State, Yaroslavl
5. Lunacharskiy AV (1958) On public education
6. Slastenin VA, Kashirin VP (2003) Psychology and pedagogy: textbook. manual for stud. of higher education Institutions. 2nd edn. Stereoplate. PC "Academy"
7. Betina OB (2006) Psychology and pedagogy: textbook. SPb
8. Kanke VA (2009) Philosophy for lawyers: textbook for students of higher educational institutions. Omega-L
9. Timofeeva EA (2012) Legal upbringing in the educational process of a law school using active and interactive forms of education. Bull Military Univ2(30)
10. Zagidulin RA (2015) Basic principles of modernization of programs and methods of training lawyers. Probl High Educ 1

Organizational and Legal Enforcement of Digital Economy in the Russian Federation at the Present Stage

Irina F. Kolontaevskaya and Larisa E. Zavedeeva

Abstract the features of the organizational and legal regulation of the digitalization of the Russian economy at the present stage are revealed in the paper. The problems arising in the process of legal regulation of digital economic phenomena and processes are examined. It is concluded that it is necessary to develop a unified law that defines the basic conceptual and categorical apparatus of the digital economic environment, establishing the principles of the state regulation of the processes of digitalization of civil circulation, securing the legal status of the main subjects of economic legal relations in cyberspace, determining security measures when using various digital technologies, etc.

Keywords Digital economy · Regulatory framework · Digitalization · Turnover

1 Introduction

The law, as a system of state-guaranteed regulatory norms, by its purpose is aimed at establishing stability, correctness and efficiency of public relations. In the context of large-scale and widespread digitalization, the organizational and functional transformation of all subjects of social interaction becomes inevitable. The growing development of relations in a virtual environment urgently requires comprehension and definition of the legal essence of new digital phenomena and processes that need to be entered into the legal system, and develop legal mechanisms for their regulation. First of all, this applies to such a globally significant and life-supporting sphere as the economy.

The organizational and strategic direction for the development of effective mechanisms for the digitalization of economic relations in Russia was determined by the

I. F. Kolontaevskaya (✉) · L. E. Zavedeeva
Moscow Witte University, 12, 2nd Kozhukhovsky Pas., Bldg. 1, Moscow 115432, Russia
e-mail: kolont@bk.ru

L. E. Zavedeeva
e-mail: 70141523@online.muiv.ru

A. V. Semenov and I. A. Sokolov (eds.), *Sustainable Development: Society, Ecology, Economy*, Earth and Environmental Sciences Library,
https://doi.org/10.1007/978-3-030-73110-6_17

Decree of the President of the Russian Federation of May 9, 2017 No. 203 "On the Strategy for the Development of the Information Society in the Russian Federation for 2017–2030". In this subordinate normative act, in particular, the general concept of the digital economy is given as a special economic activity, which, in comparison with traditional forms of management, uses the results of processing and analyzing a large amount of data in digital form, which can significantly increase the efficiency of various types of production, services, sales, delivery of goods, etc.

Decree of the President of the Russian Federation of May 7, 2018 No. 204 "On the National Goals and Strategic Objectives for the Russian Federation Development until 2024" designated the digital economy as one of the priority federal projects, in pursuance of which the Government of the Russian Federation has developed a national program "Digital Economy of the Russian Federation", providing for the creation of "end-to-end" digital technologies, including: distributed ledger systems (blockchain), big data, artificial intelligence, neurotechnologies, quantum technologies, the Internet of Things, digitalization of property rights, tokenization of civil law objects, the industrial Internet, new production and service technologies, robotics and sensorics, unmanned vehicles, communication technologies of the fifth generation, technologies of virtual and augmented reality [3, p. 74].

The Decree of the Government of the Russian Federation of March 2, 2019 No. 234 approved the management system for the implementation of the national program "Digital Economy of the Russian Federation", within the framework of which the subproject "Regulation of the digital environment" is being implemented, the key task of which is to build a system of legal regulation of the digital economy based on flexible approach in every area of public relations. The function of realization of this federal national project is entrusted to the Ministry of Economic Development of the Russian Federation, namely to the Department of Digital Economy Development.

The legislator's prompt response to the need to establish legal incentives in the field of economic activity related to the use of modern IT technologies is evidenced by the adoption of a number of federal laws on amendments and additions to existing regulatory legal acts.

Proceeding from the position that the legal basis of economic activity is enshrined in the Civil Code of the Russian Federation, the characteristics of the current state of the paradigm of civil legal regulation of digital turnover should begin with the Civil Code of the Russian Federation, which has now been amended 'piecemeal' change on selected digital rights issues. Thus, Federal Law No. 34-FZ of March 18, 2019 determined the status of digital rights and self-executing contracts (smart contracts), as a result of which citizens and businesses obtained guarantees of respect for their rights and legitimate interests, including the fundamental right to judicial protection electronic rights. The law establishes a framework for regulation of digital rights, which makes it possible to create basic conditions for the introduction into legislation of acts regulating the issue and circulation of digital rights, and also simplifies the transactions in electronic form. At the same time, digital law turned out to be implanted into civil life without proper doctrinal clarification of its nature, basis for its emergence, establishment of the object and content of this right, which causes inevitable difficulties and contradictions in practice [2, pp. 5–10].

Federal Law No. 63-FZ of April 15, 2019, which amended the Tax Code of the Russian Federation, granted VAT privileges to Russian businesses for the export of IT services, and Federal Law No. 265-FZ of July 31, 2020 introduced optimal changes in taxation for IT companies.

The procedure for attracting investments using crowdfunding platforms is established by Federal Law No. 259-FZ of August 2, 2019.

Federal laws of December 16, 2019 No. 436-FZ and No. 439-FZ amended the legislation regulating the transition to accounting information about employees in electronic form instead of paper work books, which will not only accelerate and facilitate document flow, but will also facilitate operational efficiency of solving personnel issues in companies.

Federal Law No. 476-FZ of December 27, 2019 made significant changes to Federal Law No. 63-FZ of April 6, 2011 "On Electronic Signatures". The innovations expanded the list of requirements for the accreditation of certification centers, as well as established new rules in the system for issuing qualified certificates. To ensure the credibility in the exchange of information and documents in electronic format, Federal Law No. 476-FZ introduced the institution of trusted third parties—TTP. These are organizations authorized to verify electronic signatures in electronic documents at a specific point in time and document the results of such verification.

Federal Law No. 476-FZ dated December 27, 2019 secured the transition to a registry model for the provision of state services for licensing certain types of activities, and Federal Law No. 496-FZ dated December 27, 2019 established the priority of electronic registration of the results of work (services) in the field ensuring the uniformity of measurements.

In the Russian Federation, from January 1, 2021, the Federal Law of July 31, 2020 No. 259-FZ "On Digital Financial Assets, Digital Currency and on Amendments to Certain Legislative Acts of the Russian Federation" enters into force, which significantly changes the current legal regime for the use of cryptocurrencies and blockchain in the Russian Federation. It should be noted that the Law is not devoid of contradictions. Thus, the legal definition of cryptocurrency (digital currency), set out in part 3 of Art. 1 of Federal Law No. 259-FZ, suggests its use as a means of payment, however, the same Law prohibits its use in Russia to pay for goods and services (part 5 of Art.14). Also, advertising of digital money payment methods falls within the prohibition (part 7 of article 14).

On July 31, 2020, Russian President V.V. Putin signed the law "On Experimental Legal Regimes in the Field of Digital Innovation in the Russian Federation" developed by the Ministry of Economic Development of the Russian Federation. In accordance with article 20 of the Law, this normative legal act enters into force upon the expiration of 180 days after for official publication. The law is aimed at creating legal conditions for the accelerated development and implementation of new products and services in the areas of digital innovation in all spheres of the Russian economy. At the same time, the spheres of medicine, transport, agriculture, the financial sector, education, construction and industry, and public administration were identified as the key areas for testing technologies.

The legislator has determined that the program of the experimental legal regime is an act of special regulation establishing special regulation for a specific experimental legal regime. At the same time, such regulation applies only to participants in the "regulatory sandbox" with strong restrictions in time and territory. The establishment of the experimental regime is preceded by a number of mandatory procedures with the involvement of sectoral authorities in order to comprehensively assess the possible risks of conducting a legal experiment. The final decision on the possibility of conducting a legal experiment is made by the Government.

As the study showed, a number of key regulatory legal acts have been adopted recently aimed at regulating the processes of using various technologies and tools of the digital economy.

At the same time, digital phenomena and processes have their own specifics of legal registration and regulation, and haste in resolving new digital economic realities is fraught with great risks and financial losses. The development of the digital economy to a decisive extent depends on the quality of the draft laws being developed and adopted. Legal rule-making should be preceded by an understanding of the legal nature and legal essence, as well as an understanding of the legal value of those social relations that are subject to legalization. At the same time, virtual objects must be legally included in civil circulation, thereby ensuring the protection of rights to them. Depending on the purpose and potential of information technology, the conditions of its use and the expected result, it is necessary to apply various combinations of legal measures, including judicial practice and established traditions of business transaction.

It seems advisable to adopt a single law that would define in its norms the basic conceptual and categorical apparatus of the digital economic environment, establish the principles of state regulation of the digitalization of civil circulation, consolidate the legal status of the main subjects of economic legal relations in cyberspace, determine security measures when using various digital technologies, etc. At the same time, according to experts, it is currently impossible to develop a systemic federal law that provides a unified full-fledged framework for the comprehensive regulation of new digital relations. At the same time, regulation cannot be taken to the subordinate level [3]. Therefore, at the moment, there is a point introduction of amendments to the current legislation, and concretization is carried out at the level of subordinate normative legal acts.

Thus, we can conclude that the current virtual economic environment, along with its high adaptability and extreme diversity of its constituent elements, is characterized by the spontaneity of the emergence and spread of new digital realities, the unpredictability of side effects and risks, legislative inconsistency and uncertainty in the regulation of individual processes of digitalization of civil circulation. Solving the problem of introducing the digital economy into an expedient and harmonious legal field will help regulate and stimulate the development of this area, which is now an important direction of the legislative policy of the Russian state.

References

1. Avakyan SA, Barenboym PD, Komarova VV, Barenboym PD (2018) Yearbook of constitutional economics. LUM, 440 pp
2. Gorodov OA (2020) Acquisition of utilitarian digital rights as a new way of investing under Russian legislation. Law Digit Econ 1:5–10
3. Belitskaya AV, Belykh VS, Belyaeva OA (2019) Legal regulation of economic relations in modern conditions of digital economy development, Vaipan VA, Egorova MA (eds) Justitsinform, 376 pp

Corruption Parameters

Aleksandra V. Krekhovets, Marina A. Stepanova, and Evgeny V. Tsarev

Abstract The purpose of this article is to study the parameters of the socio-legal phenomenon of "corruption" in certain power structures and the effectiveness of using it in the fight against corruption crime. "Corruption" is a social phenomenon that causes serious threats to the security and sustainability of the functioning of any society. In modern society, corruption is diverse in its manifestations. The essence of corruption, its types and parameters are considered in the article. The methodological basis of the study was the dialectical method and other general scientific methods of cognition: analysis, the system method, and also particular scientific methods: formal-logical, generalization and study of the results of research conducted by the INDEM Foundation. The main parameters of corruption are identified, and the author's formulation of sociological parameters of corruption is formulated as the most significant for the study of socio-legal phenomenon.

Keywords Corruption · INDEM foundation · Corruption parameters · Corruption perception index · Sociological parameters

1 Introduction

The phenomenon of corruption is well known for a long time and is perceived as a given of all states (countries of the world) in any periods of its development. The essence of corruption as a social and legal phenomenon, its causes and consequences,

A. V. Krekhovets (✉) · M. A. Stepanova
Nizhniy Novgorod Academy of the Ministry of the Interior of Russia, Ankudinovskoe Sh 3, Nizhny Novgorod 603950, Russia
e-mail: alexa-xa@mail.ru

M. A. Stepanova
e-mail: marina-stepanova-1979@bk.ru

E. V. Tsarev
Moscow Witte University, Branch in N.Novgorod, 51 Bolshaya Pecherskaya Str, Nizhny Novgorod 603155, Russia
e-mail: tsarevzhenya@yandex.ru

A. V. Semenov and I. A. Sokolov (eds.), *Sustainable Development: Society, Ecology, Economy*, Earth and Environmental Sciences Library,
https://doi.org/10.1007/978-3-030-73110-6_18

anti-corruption measures are the subject of unabated disputes in our country. But the most significant problem in our opinion is the determination of the parameters of corruption. Its connected to the fact that now there is no single approach to determining the level of corruption in various countries of the world. Despite this, a sociological research is almost always carried out, based on an analysis of the perception of corruption by the population of the country.

The determining the parameters of corruption is a prerequisite for conducting an adequate study of this socio-legal phenomenon. The validity of its result will depend on what parameters will be used as the basis for studying corruption. In this regard, within the framework of this article we will try to examine the nature of the corruption origin, its parameters in order to determine the scope of such a socio-legal phenomenon as corruption in certain power structures and how effectively it can be used to combat corruption and crime in general.

2 Corruption Perception Index for the Last Nine Years

Corruption is a complex social phenomenon of all countries. As a social phenomenon, corruption is part of the social reality understood as the objectified result of people's subjective activity. Once created, this reality begins to develop and function according to its own objective laws, exerting the opposite effect on the people who created it [10, p. 96].

The President of the Foundation INDEM G. A. Satarov correctly noted: "The growth of corruption in Russia in recent years has been causing worry in society. The danger of this phenomenon for the further development of our country was repeatedly mentioned at the highest level. Although the population regularly learns from the media about the successes in fighting it, one cannot say that victory is near. Moreover, as the results of research conducted by the INDEM Foundation show, the situation gets worse every year" [12, p. 4].

It should be noted that the Corruption Perceptions Index (CPI), compiled by Transparency International, firstly measures the level of perception of corruption in the public sector of a country; secondly, is a composite index based on surveys conducted among experts and in business circles.

So, according the CPI in 2009, Russia scored 2.2 points and ranked 146 out of 180 countries.

In 2010, Russia scored 2.1 points and ranked 154 out of 178 places. Conclusion: The level of corruption in Russia requires immediate actions in practice.

In 2011–2.4 points, ranked 143 out of 183. Conclusion: There is a positive trend, which is explained by the adoption of a complex of anti-corruption laws in Russia.

In 2012, the CPI was measured according to the methodology (scale from 0 to 100 points, where zero indicates the highest level of perception of corruption, and a hundred—the lowest). So, according the CPI in 2012, Russia scored 28 points and ranked 133 out of 176 countries.

In 2013, Russia received 28 points, which corresponds to 127th place out of 177 countries. Conclusion: The measures taken by the authorities turned out to be insufficient to change the level of corruption in Russia in a positive direction.

In 2014, Russia received 27 points (one point less than in 2013) and ranked 136th position. Conclusion: Because of the indecision in the prosecution of corruption crimes and the randomness of measures to prevent corruption, Russia remains in the last third of the CPI.

In 2015, Russia took the 119th place (29 points). Conclusion: Changing the place of Russia in the CPI is dictated by the introduction into the legal field of a number of restrictions that are very burdensome for public officials in terms of declaring property and income, as well as owning foreign property.

In 2016, Russia scored 29 points out of 100. Conclusion: The position of Russia has not actually changed, it received the same number of points, and the decrease in position in the rating (from 119th to 131st) is due to the fact that there are more countries in it this year.

In 2017, Russia took 135th place (29 points) out of 180. Conclusion: For the third consecutive year, Russia is gaining 29 points in the CPI. A whole series of high-profile corruption cases in Russia did not make enough impression on respondents to recognize any progress in countering corruption.

In 2018, Russia ranked 138 out of 180 and scored 28 points out of 100. Conclusion: For the last three years, Russia scored 29 points, and this year it has lost one point and dropped three places [5].

3 Types of Corruption

In modern criminological science, on the basis of various criteria, there are different directions for studying corruption, and types of corruption.

1. By initiator of corrupt relations:
    - "corruption from below"—corruption initiated by employees in relation to heads;
    - "corruption from above"—corruption initiated by heads in relation to employees [7, p. 10], [4, c. 39–40], [13, p. 67].
2. On the main functions implemented by state and municipal bodies:
    - corruption in the executive branch [16, pp. 153–189];
    - corruption in the legislative bodies of state power [6];
    - corruption in the judiciary [13, pp. 8–9].
3. According to the territorial prevalence and national composition of participants in corruption relations:

- transnational (international)—corruption, directed to monopolizing relations in a particular area within the global community;
- intrastate—corruption that has become widespread on the territory of individual states or in specific national societies [13, pp. 8–9].

4. According to the depth of the defeat of the legal consciousness of the ruling political elite and the form of manifestation of corruption in modern Russian society:

   - "light corruption"—corruption, which is not connected to violation of duties in the service;
   - "gray corruption"—corruption connected to the commission of offenses in the service;
   - "black corruption" is corruption that is not connected to the commission of official crimes or corruption that is connected to complicity in organized criminal activity [1, p. 98].

5. By the degree of activity:

   - "active corruption"—corruption which is not connected active bribery;
   - "passive corruption"—corruption which is not connected passive bribery [2, p. 118].

6. On the sphere of corruption inside the society:

   - domestic corruption—an extortion by official money or material values from the population;
   - business corruption—a transfer money by businessmen to officials;
   - administrative corruption—extortion money or property by officials from businessmen;
   - corruption in the form of "state capture"—a buying power decisions by businessmen in their own interests or the interests of their enterprises;
   - corruption in the form of "business capture"—an establishment by officials the control over firms in order to obtain an additional source of income [7, p. 12].

7. By field of activity:

   - corruption in public administration;
   - parliamentary corruption;
   - corruption in enterprises [9].

4 The Concept of "Parameters" and "Sociological Parameters" of Corruption

In the explanatory dictionaries of the Russian language there are different meanings of the word "parameter". So, referring to the legal dictionary, the term "parameter", which is translated from the Greek. parametron—measuring, means: firstly, the indicator, the value of which remains constant within the limits of the task; secondly, the value, the indicator characterizing any property of the device, process, substance; and, thirdly, an indicator whose value has a significant impact on the condition of the economic process [3, p. 781].

So Kotlyarov S. B., Chicherov E. A. as the main parameters of measurement or indicators of corruption emphasize the following: the number of facts of participation in corruption relations (identified by means of surveys or interviews); frequency of participation in corrupt relations (intensity); the average size of a bribe in a state, region, city, etc.; the amount of money spent in the course of a corruption relationship by an individual or organization; risk of involvement in corruption; the perception of the level of corruption in society, the number of corruption offenses in the official statistics of the internal affairs bodies; the number of corruption scandals in the media; the number of corruption violations in public procurement [8, p. 9].

Sidorenko E. L. adds to the above-mentioned new criminological parameters for measuring corruption in Russia: the disciplinary corruption index (the results of prosecutorial oversight over the implementation of anti-corruption legislation and the dynamics of disciplinary liability for violation of anti-corruption legislation); corruption incidence index (the ratio of the number of persons identified during a certain time in connection with receiving a bribe (commercial bribery) to the total number of employees of this body or organization); typicality of corruption (the ratio of the types of detected crimes to the structure of crimes of corruption in general); geography of corruption; transaction costs (the ratio of the benefits and costs of corrupt behavior) [14].

It should be noted that the need to study corruption is dictated, firstly, by the tasks of developing an anti-corruption policy. Systematic measurements of corruption provide the basis for its monitoring, which allows assessing not only the current state, but also changing various parameters characterizing this phenomenon in order to take adequate measures to combat this phenomenon [11, p. 8].

The following methods are used to measure corruption:

- analysis of criminal statistics, analysis the information of control bodies, legislation, the press;
- institutional analysis;
- sociological methods (surveys, questionnaires, interviews, focus groups).

Based on the method and scope of the study, the following groups of indicators of corruption can be distinguished:

1. official statistics;
2. media indicators on corruption;

3. indicators of anti-corruption legislation.

5 Sociological Indicators of Corruption

In our opinion, analyzing the state of corruption, it is necessary to focus not on official statistics, but on the results of sociological research, but in general, the main trends associated with the detection of corruption crime and the public's view of it should coincide, that is, a simultaneous increase or decrease in indicators. If these indicators are multidirectional, then you need to focus on the indicators of sociological research.

In this regard, it is necessary to investigate the sociological parameters of corruption, which should be understood as an indicator, a quantity characterizing corruption as a socio-legal phenomenon, determining its existence and revealing its social essence, as well as having a significant impact on the boundaries of the spread of corruption in certain power structures.

6 Study of Sociological Parameters of Corruption

In our opinion, it is advisable to conduct a study of the sociological parameters of corruption, by example, domestic and business corruption. Today, these types of corruption are most common in society.

Let's calculate the parameters of corruption using the method of a sociological survey of citizens (domestic corruption) and a survey of entrepreneurs (business corruption). Sociological survey is the most common method of sociology, which is used to obtain primary information, characterized by efficiency, simplicity and cost-effectiveness. It makes it popular among entrepreneurs and managers, public and political organizations.

Based on a survey of these groups, the parameters of corruption can be calculated, including quantitative and qualitative indicators that are to be investigated. Surveys are conducted in the form of a questionnaire and are aimed at studying domestic and business corruption, taking into account the opinions of citizens and entrepreneurs on the state of corruption in a particular region and on the most effective measures to combat it.

1. Using the results of citizens survey (domestic corruption), the following sociological parameters (indicators) take place.

    1.1. Quantitative indicators are:

    - coverage of corruption, the proportion of citizens who at least once became participants in the corruption situation;
    - willingness to corruption, the proportion of citizens who are willing to commit a corrupt transaction;
    - the average size of a bribe, according to citizens;

- the share of bribes in the annual budget of citizens.

1.2. Quality indicators are:

- the level of corruption in a particular region and in Russia as a whole;
- the level of corruption in individual enterprises and institutions;
- ranking of organizations by corruption, according to citizens;
- assessment of the dynamics of corruption in a particular region;
- preferences of citizens regarding individual anti-corruption measures.

2. Using the results of entrepreneurs survey (business corruption).

2.1. Quantitative indicators are:

- Coverage of corruption, the share of entrepreneurs who at least once became participants in the corruption situation;
- willingness to corruption, the proportion of organizations that are willing to commit a corrupt transaction;
- the share of respondents in whose practice there were cases of informal resolution of issues with a particular authority;
- the share of respondents who knows the facts of informal resolution of issues with a particular authority with the participation of their competitors, partners, and acquaintances.

6.1 Quality Indicators Are

- the most significant obstacles to the operation of a business in a particular region;
- the level of corruption in the bodies and institutions of a particular region;
- rating of bodies and institutions of corruption, in the opinion of entrepreneurs;
- the main causes of corruption in the bodies and institutions in a particular region;
- the degree of interest of entrepreneurs in corruption;
- the reasons why businessmen make a corrupt deal;
- the effectiveness of anti-corruption measures in a particular region, in the opinion of entrepreneurs;
- changing the level of corruption in a particular region;
- preferences of entrepreneurs regarding particular anti-corruption measures.

7 Conclusion

Based on the above, we can make the following conclusions.

First, to determine the limits of the spread of corruption as a socio-legal phenomenon, it is necessary to determine its parameters and first of all sociological ones.

Secondly, the sociological parameters of corruption is an indicator, a quantity characterizing corruption as a socio-legal phenomenon, determining its existence and revealing its social essence, as well as having a significant impact on the boundaries of the spread of corruption in certain power structures.

Thirdly, in order to calculate sociological parameters of corruption (quantitative and qualitative indicators), it is necessary to conduct sociological polls (in the form of a questionnaire) of certain categories of citizens, in order to obtain primary information and study corruption, taking into account the opinions of citizens about the state of corruption in a particular region and about most effective interventions to combat it.

Fourth, the results of surveys of certain categories of citizens allow us to reveal the opinion of citizens and specialists about the corruption of a society, are not fully reliable, as they are subject to the influence of subjective factors.

Fifth, the analysis of the existing parameters of corruption in practice is debatable, which directly indicates the lack of currently accepted methodologies that provide reliable data on corruption. In this regard, it seems necessary to conduct an interdisciplinary study using the proposed sociological methodology for measuring corruption and investigating corrupt behavior, and a comprehensive system of measures aimed at reducing corruption in the Russian Federation, taking into account the specifics of Russian society, can be proposed.

References

1. Baranov VM (2002) Shadow Law: Monograph. Nizhny Novgorod Academy of the Ministry of Internal Affairs of Russia, N. Novgorod
2. Beken TV (2002) European view on the Russian draft federal law “On the fundamentals of anti-corruption policy”. State and law, № 6
3. Kuznetsov SA (2000) (ed) Big explanatory dictionary of the Russian language, Ch. Norint, SPb
4. Voluev VN, Makarycheva AS (2000) Corruption in Russia: political approaches. Corruption in Russia: municipal, regional, federal and international aspects: Analytical report. N. Novgorod
5. Grib VG, Oks LE (2019) Counteracting the corruption: study guide. In: Grib VG, Oks LE (eds) Moscow Academy of Finance and Industry, 2011; Corruption Perception Index 2010–2018. https://www.transparency.org. Accessed 26 June 2019
6. Danilov SU (1992) Legislative measures against parliamentary corruption in Canada. USA. Economy. Politics. Ideology, № 5
7. Kabanov PA, Gazimzyanov RR (2003) Corruption in Russia: concept, essence, causes, counteraction: Tutorial. Naberezhnye Chelny: Publishing House “Strezhen”
8. Kotlyarov SB, Chicherov E (2015) A. Indicators for measuring corruption: the concept and improvement [Text]. Actual problems of law: Materials of the IV Intern. scientific conf. (Moscow, November 2015). Buki-Vedi, pp 6–9. https://moluch.ru/conf/law/archive/179/9057/. Accessed 26 June 2019
9. Kuznetsov YuA, Silinsky YuR, Khomutova AV. Russian and Foreign Legislation on Anti-Corruption Measures: A Study Guide for Universities Institute of Legislation and Comparative Law under the Government of the Russian Federation [Electronic resource]. - Access mode: http://www.anticrimecom.org/cgi-bin/page.cgi?id=213
10. Osipov GV (2010) Introduction to sociological science

11. Osipov GV (2012) Report on the research work "Development of scientifically grounded sociological parameters of corruption". RAS Institute of Socio-Political Studies, RAS
12. Satarov GA (2007) How to measure and control corruption. Questions of economy. № 1
13. Selikhov NV (2001) Corruption in the state mechanism of modern Russia (theoretical aspects): Author's abstract. dis ... cand. legal sciences. Ekaterinburg
14. Sidorenko EL (2019) New criminological parameters for measuring corruption in Russia. Bull Moscow Univ Ministry Internal Affairs of Russia №4. https://cyberleninka.ru/article/n/novye-kriminologicheskie-parametry-izmereniya-korruptsii-v-rossii (appeal date: 26.06.2019). CyberLeninka: https://cyberleninka.ru/ article/ n/ novye-kriminologicheskie-parametry-izmereniya-korruptsii-v-rossii
15. Eksanova AA (2001) Bribery as a crime: legal nature, qualifications and responsibility: Monograph, Pn Panchenko (ed). N. Novgorod
16. Estrin AYa (1913) Bribery in legislation and doctrine. Transactions of the criminal law circle of St. Petersburg University. SPb

Printed by Printforce, the Netherlands